BRAIN LIPIDS AND DISORDERS IN BIOLOGICAL PSYCHIATRY

New Comprehensive Biochemistry

Volume 35

General Editor

G. BERNARDI
Paris

ELSEVIER

Amsterdam · London · New York · Oxford · Paris · Shannon · Tokyo

Brain Lipids and Disorders in Biological Psychiatry

Editor

E. Roy Skinner

*Department of Molecular and Cell Biology,
University of Aberdeen, Aberdeen, Scotland, UK*

2002

ELSEVIER

Amsterdam · London · New York · Oxford · Paris · Shannon · Tokyo

ELSEVIER SCIENCE B.V.
Sara Burgerhartstraat 25
P.O. Box 211, 1000 AE Amsterdam, The Netherlands
© 2002 Elsevier Science B.V. All rights reserved.

First edition 2002

Library of Congress Cataloging in Publication Data
A catalog record from the Library of Congress has been applied for.

ISBN: 0-444-50922-4
ISBN: 0-444-80303-3 (Series)
ISSN: 0167-7306 (Series)

⊗ The paper used in this publication meets the requirements of ANSI/NISO Z39.48-1992 (Permanence of Paper).
Printed in The Netherlands.

Preface

More than half of the mass of the brain is lipid and, unlike most other tissues of the body, the major proportion of the fatty acids associated with this lipid consists of specific long-chain polyunsaturated fatty acids. It is a remarkable fact that these components and their precursor fatty acids which are vital for the normal functioning of the brain, and yet cannot be synthesised in the human body, tend to be in short and often inadequate supply in the western diet; this situation is further frustrated by the slow rate of conversion of the precursors into these fatty acids in a number of situations.

It is therefore not surprising that deficiences in the uptake or defects in the metabolism of these constituents should be associated with impaired brain function; there is, in fact, good evidence to suggest that this may contribute to retarded mental development in early life, behavioral abnormalities and mental disorders including schizophrenia and depression.

Likewise, alterations in the recycling and turnover of cholesterol and phospholipids, major constituents of brain cell membranes, are also suggested to exert an influence on brain function and may contribute, in part, to several conditions including Alzheimer's disease and personality disorders and violence.

A knowledge of the possible roles of lipids in the development of these conditions is therefore of the utmost importance as a better understanding of the underlying causes of the latter may contribute towards a means of prevention, amelioration and cure of these most debilitating of medical disorders. The need for such investigations is particularly timely in view of the rapid increase in the incidence of many of these disorders and the prediction that mental disorders and particularly depression will be the major illness and the greatest demand on health service resources in the next decade (WHO Report).

This volume presents a collection of chapters which provide evidence, including that from very recent studies, for an association between brain lipids and disorders in biological psychiatry, contributed by leading authorities in their respective fields.

I am most grateful to all of the authors for their contributions to this volume which I hope will stimulate further interest and lead to increased research in this important development area. It is also a pleasure to acknowledge the considerable advice and encouragement given to me throughout the preparation of the volume by Dr Frank Corrigan, especially with regard to the design of the book; without his guidance, the book would have lacked much of its character.

E Roy Skinner
Aberdeen
February 2001

List of contributors*

Gregory J. Anderson   23
*The Division of Endocrinology, Diabetes and Clinical Nutrition,
Department of Medicine, L465, Oregon Health Sciences University,
Portland, OR 97201-3098, USA*

Graham C. Burdge   159
Institute of Human Nutrition, University of Southampton, Southampton, UK

William E. Connor   23
*Division of Endocrinology, Diabetes and Clinical Nutrition,
Department of Medicine, L465, Oregon Health Sciences University,
Portland, OR 97201-3098, USA*

Frank M. Corrigan   113
Argyll and Bute Hospital, Lochgilphead, Argyll PA31 8LD, Scotland, UK

Marc Danik   53
*Douglas Hospital Research Centre, Faculty of Medicine, McGill University,
Montréal, Québec, Canada*

Akhlaq A. Farooqui   147
*Department of Medical Biochemistry, The Ohio State University, 1645 Neil Avenue,
Columbus, OH 43210, USA*

Tahira Farooqui   147
*Department of Medical Biochemistry, The Ohio State University, 1645 Neil Avenue,
Columbus, OH 43210, USA*

J. Stewart Forsyth   129
Department of Child Health, University of Dundee, Dundee, DD1 9SY, Scotland, UK

Joseph R. Hibbeln   67
*Chief, Outpatients Clinic, National Institute of Alcohol Abuse and Alcoholism,
Park 5, Room 158, MCS 8115, Bethesda, MD 20852, USA*

David Horrobin   39
*Laxdale Ltd, Kings Park House, Laurelhill Business Park, Polmaise Road,
Stirling FK7 9JQ, Scotland, UK*

---

* Authors' names are followed by the starting page number(s) of their contribution(s).

Lloyd A. Horrocks   147
Department of Medical Biochemistry, The Ohio State University, 1645 Neil Avenue, Columbus, OH 43210, USA

David J. Kyle   1
Martek Biosciences Corp, 6480 Dobbin Rd., Columbia, MD 21045, USA

Kevin K. Makino   67
Outpatients Clinic, National Institute of Alcohol Abuse and Alcoholism, Park 5, Room 158, MCS 8115, Bethesda, MD 10852, USA

Judes Poirier   53
Douglas Hospital Research Centre and McGill Centre for Studies in Aging, Department of Psychiatry, Neurology and Neurosurgery, McGill University, Montréal, Québec, Canada

Anthony D. Postle   159
Department of Child Health, University of Southampton, Level G (803), Centre Block, Southampton General Hospital, Tremona Road, Southampton, SO16 6YD, UK

E. Roy Skinner   113
Department of Molecular and Cell Biology, University of Aberdeen, MacRobert Building, Room 809, Regent Walk, Aberdeen AB24 3FX, Scotland, UK

Peter Willatts   129
Department of Psychology, University of Dundee, Dundee, DD1 4HN, Scotland, UK

Contents

Chapter 7. Do long-chain polyunsaturated fatty acids influence infant cognitive behavior?
J.S. Forsyth and P. Willatts

Chapter 8. Molecular species of phospholipids during brain development. Their occurrence, separation, and roles
Akhlaq A. Farooqui, Tahira Farooqui and Lloyd A. Horrocks

Other volumes in the series

E.R. Skinner (Ed.), *Brain Lipids and Disorders in Biological Psychiatry*

CHAPTER 1

The role of docosahexaenoic acid in the evolution and function of the human brain

David J. Kyle

*Martek Biosciences Corp, 6480 Dobbin Rd.,
Columbia, MD 21045, USA, tel.: (410)-740-0081; fax: (410)-740-2985*

1. Introduction

Docosahexaenoic acid (DHA) is a primary building block of the membranes of the human brain and visual system. It is a highly unsaturated, very long chain polyunsaturated fatty acid (PUFA) which is metabolically expensive to make and maintain as a membrane component. DHA exhibits unique conformational characteristics that allow it to carry out a functional as well as a structural role in biological membranes of high electrical activity. The structural role involves an intimate association with certain membrane proteins such as those, which have a specific seven-transmembrane spanning structural motif. Included in these are the G-protein coupled receptors and certain ion conductance proteins, which have critically important functions in cell signalling and metabolic regulation. One functional role suggested for DHA involves specific control of calcium channels by the free fatty acid, thereby representing an endogenous cellular control mechanism for maintaining calcium homeostasis.

DHA is an ancient molecule. It has been exquisitely selected by nature to be a component of visual receptors and electrical membranes in various biological systems for 600 million years. It is found in simple marine microalgae (Behrens and Kyle 1996), in the giant axons of cephalopods, and in the central nervous system and retina of all vertebrates (Bazan and Scott 1990, Salem et al. 1986). Indeed, in mammals it represents as much as 25% of the fatty acid moieties of the phospholipids of the grey matter of the brain and over 50% of the phospholipid in the outer rod segments of the retina (Bazan et al. 1994). Aquatic environments are replete with plants, micro-organisms and animals which are highly enriched in DHA and, during the period which coincided with the movement of ancestral hominids to an ecological niche rich in DHA – the land/water interface – there was a rapid expansion of the brain to body weight ratio. It has been speculated that it was this movement to a DHA-rich environment approximately 200 000 years ago which allowed the rapid development of the big brain of *Homo sapiens* (Broadhurst et al. 1998). Supporting this concept, recent fossil discoveries have provided evidence that hominids of 120 000 to 80 000 years ago were making extensive use of the marine food chain.

As a result of its fundamental role in neurological membranes of humans, the clinical consequences of deficiencies of DHA, or hypodocosahexaenemia, range from the profound (*e.g.*, adrenoleukodystrophy) (Martinez 1990) to the subtle (*e.g.*, reduced night vision) (Stordy 1995). DHA also plays a key role in brain development in humans. A specific DHA-binding protein expressed by the glial cells during the early stages of brain development, for example, is required for the proper migration of the neurons from the ventricles to the cortical plate (Xu et al. 1996). DHA itself is concentrated in the neurites and nerve growth cones and acts synergistically with nerve growth factor in the migration of progenitor cells during early neurogenesis (Ikemoto et al. 1997).

The pivotal role of DHA in the development and maintenance of the central nervous system has major implications to adults as well as infants. The newly recognized, multifunctional roles of DHA may serve to explain the long-term outcome differences between breast-fed infants (getting adequate DHA from their mother's milk) and infants who are fed formulas which do not contain supplemental DHA (Anderson et al. 1999, Crawford et al. 1998). In summary, DHA is a unique molecule, which is critical to normal neurological and visual function in humans, and we need to ensure that we obtain enough of it in the diet from infancy to old age.

2. The universality of DHA in neurological tissues

2.1. The building block of the neuro-visual axis

Although DHA is found throughout the body, the highest concentrations are found in the membranes of neurological tissues. Indeed, it is the most abundant PUFA comprising the phospholipid of the grey matter of the brain, representing over 30% of the fatty acids of the phosphatidyl ethanolamine (PE) and phosphatidyl serine (PS) in the neuron (Salem et al. 1986). It is primarily concentrated in the synaptic plates and synaptosomes, but is generally not found in high abundance in the myelin sheath. DHA is also associated with the neurite growth cones where it has been shown to promote neurite outgrowth (Martin 1998). Concomitant with these high tissue concentrations, the observed functional roles of DHA (Fig. 1) are primarily associated with tissues of high electrical activity.

The vertebrate retina is an extension of the neurological tissues and also contains a very high level of DHA. The DHA is found in the highest concentration in the retina associated with rhodopsin, the light transducing protein. This protein has a seven transmembrane (7-TM) structural motif and is found in the layered membrane of the rod and cone outer segments (Fig. 2). The DHA content of these membranes is the richest in the body and there is a complex recycling mechanism that maintains these high levels in spite of a very rapid turnover of the membranes themselves (Bazan and Rodriguez de Turco 1994). DHA represents over 50% of the fatty acid moieties of the PE found in the rod outer segments and many of these phospholipids are present as di-DHA phospholipids. During excitation, the outer layers of the photoreceptor are shed, phagocytozed, and the membranes are recycled to form new disks at the opposite end of the disk stack (Rodriguez de Turco et al. 1997).

2.2. The biosynthesis of DHA

All fatty acids are synthesised by a biochemical mechanism involving the successive extension of precursor molecules by 2-carbon increments. Unsaturated fatty acids are the products of desaturase enzymes that act at specific positions in the acyl chain. Insertion

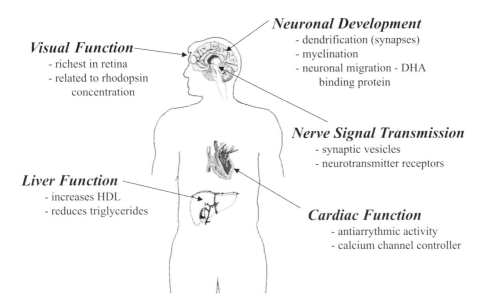

Visual Function
- richest in retina
- related to rhodopsin
  concentration

Neuronal Development
- dendrification (synapses)
- myelination
- neuronal migration - DHA
  binding protein

Nerve Signal Transmission
- synaptic vesicles
- neurotransmitter receptors

Liver Function
- increases HDL
- reduces triglycerides

Cardiac Function
- antiarrythmic activity
- calcium channel controller

Fig. 1. Overview of the functional roles of DHA in the human body.

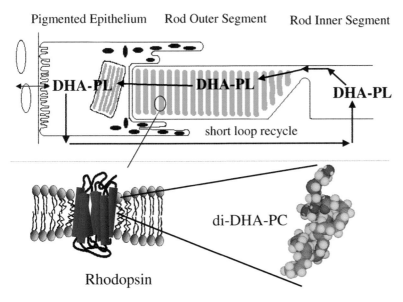

Pigmented Epithelium   Rod Outer Segment   Rod Inner Segment

DHA-PL   DHA-PL   DHA-PL

short loop recycle

di-DHA-PC

Rhodopsin

Fig. 2. Diagrammatic representation of the outer segment of a rod cell with the rhodopsin- and DHA-rich membrane stacks. Expanded view represents the 7TM rhodopsin situated in a lipid bilayer with DHA-rich phospholipids in the boundary region around the protein.

of double bonds at positions 12 and 15 of oleic acid, for example, requires $\Delta12$- and $\Delta15$-desaturases, respectively. Such desaturases which act near the methyl end of the acyl chain are present in plants but not in mammals. Consequently, humans cannot synthesize

4

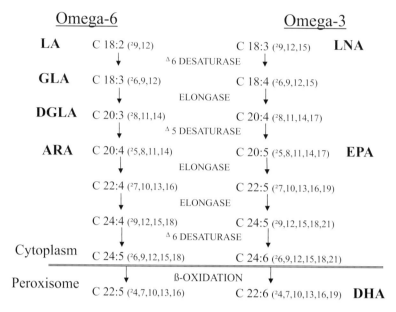

Fig. 3. The omega-6 and omega-3 fatty acid biosynthetic pathways including the final steps of DHA synthesis which take place in the peroxisome.

linoleic acid (LA, or C18:2 (Δ9,12)) or linolenic acid (LNA; or C18:3 (Δ9,12,15)) *de novo*, and so these are referred to as the essential fatty acids. LA and LNA are readily obtained from plants in our diet and are the parent molecules of the omega-6 and the omega-3 families of fatty acids which include arachidonic acid (ARA) and DHA, respectively (Fig. 3). It has been well documented, that most mammals have the ability to synthesise all the fatty acids of the omega-3 and omega-6 pathways from these two precursors (Greiner et al. 1997, Salem et al. 1996). Some species (*e.g.*, cats and other obligate carnivores) cannot synthesise DHA or ARA at all, and these fatty acids are essential components in their diets (Salem and Pawlosky 1994). Mice and other vegetarian species, on the other hand, are quite efficient at synthesising ARA and DHA from their dietary precursors of LA and LNA. Humans are somewhat intermediate, and the fractional synthetic rate is likely to be too slow to keep up with the requirements of the rapidly developing brain during the last trimester *in utero* and the first few years of post-natal life (Cunnane et al. 1999, Salem et al. 1996). Under these conditions (*e.g.*, infancy), DHA and ARA have been referred to as "conditionally essential" in humans (Carlson et al. 1994). This may also be the case in certain clinical pathologies and old age.

The final step in the synthesis of DHA was believed to involve a Δ4-desaturation of the omega-3 docosapentaenoic acid (C22:5 Δ7,10,13,16,19)(DPA), to produce DHA (C22:6 Δ4,7,10,13,16,19). After a long search for the elusive Δ4-desaturase, Sprecher and colleagues concluded that the synthesis of DHA actually involved a more elaborate pathway (Voss et al. 1991). The omega-3 DPA (C22:5) is elongated to 24:5(Δ9,12,15,18,21), and then desaturated with a Δ6-desaturase to form 24:6(Δ6,9,12,15,18,21). This fatty acid is then transferred to the peroxisome where it undergoes one cycle of beta-oxidation to form DHA (Fig. 3). A similar process can occur with the omega-6 pathway to form 22:5(Δ4,7,10,13,16), docosapentaenoic acid (n-6), when an animal is forced into a severe omega-3 fatty acid deficiency (Green et al. 1997).

2.3. The role of DHA in membrane structure

The twenty carbon fatty acids of the omega-6 and omega-3 families such as ARA and eicosapentaenoic acid (EPA) are the precursors for a family of circulating bioactive molecules called eicosanoids. DHA, with twenty-two carbons however, is not an eicosanoid precursor. On the other hand, DHA is the most abundant omega-3 fatty acid of the membrane phospholipids that make up the grey matter of the brain. More specifically, it is found in exceptionally high levels in the phospholipids that comprise the membranes of synaptic vesicles (Arbuckle and Innis 1993, Bazan and Scott 1990, Wei et al. 1987). It is preferentially taken up into the brain and synaptosomes (Suzuki et al. 1997) and is also found in high concentrations in the retina (Bazan and Scott 1990), cardiac muscle (Gudbjarnason et al. 1978), and certain reproductive tissues (Connor et al. 1995, Conquer et al. 1999). DHA, therefore, is thought to play a unique role in the integrity or functionality of all of these tissues (Fig. 1).

Biological membranes are comprised of a phospholipid matrix and membrane proteins. Both components play an important role in conferring specificity to that particular membrane. For example, in the outer segments of the rod and cone cells of the retina there are a series of pancake-like stacks of membranes in which the retinal-binding protein, rhodopsin, is found (Fig. 2) (Bazan and Rodriguez de Turco 1994, Gordon and Bazan 1990). Clandinin and colleagues recently demonstrated that the concentration of rhodopsin in these membranes was dependent on the DHA content of the phospholipids comprising those membranes (Suh et al. 1996). That is, the greater the DHA content, the higher was the rhodopsin density (a characteristic that should define the light sensitivity of the eye). On the other hand, if an animal is made omega-3 deficient and the DHA in the retina is replaced its omega-6 counterpart (omega-6 docosapentaenoic acid – DPA; C22:5 Δ4,7,10,13,16), the visual processing of the animal is compromised (Neuringer et al. 1986). Since the existence of an additional double bond at the Δ19 position of an otherwise identical molecule, has such a dramatic effect on the performance of a specific organ (Bloom et al. 1999, Salem and Niebylski 1995), it is believed that DHA must play a pivotal role in the "boundary lipid" of rhodopsin such that any alteration in the boundary lipid results in a significant impact of its biochemical function (Fig. 2).

It appears that DHA is also found in close association with many other membrane proteins of the 7-transmembrane motif (7-Tm) in addition to rhodopsin. These include the G-protein coupled receptors which represent important cell signalling mechanisms. Most of the neurotransmitter receptors also have a similar protein structural motif and high concentrations of DHA are found in the membranes at the synaptic junction between neurons. The observation that DHA levels appear to be correlated with levels of certain neurotransmitters (e.g., serotonin) may be partially explained by the requirement of certain DHA-containing phospholipids for the optimal performance of the neurotransmitter receptors (Hibbeln et al. 1998a, Higley and Linnoila 1997). There is a large literature on DHA deficits reducing memory and cognitive abilities in laboratory animals from the early work of Lamptey (Lamptey and Walker 1976) to the most recent work by Greiner (Greiner et al. 1999). Recent studies indicate that DHA also appears to be a factor in the hippocampal acetylcholine activity (Young et al. 1998) suggesting its role in the membrane facilitates signal transduction and, thereby, in cognitive function (Minami et al. 1997).

2.4. The role of DHA in membrane function

DHA may play a greater role than simply a structural element in many biological membranes. It may also modulate many important biochemical events that occur within the cell itself. DHA has been shown to be involved in the regulation of certain genes (Clarke and Jump 1994, Sesler and Ntambi 1998). The most commonly reported effect of PUFAs, in general, is a decrease in the activity of liver enzymes involved in lipogenesis, with a consequential lowering of circulating triglycerides (Fig. 1). Several clinical studies have reported consistent results in triglyceride lowering with DHA supplementation alone (Agren et al. 1996, Conquer and Holub 1996, Davidson et al. 1997, Innis and Hansen 1996). This is a reversible effect and dietary PUFA restriction can induce the expression of lipogenic enzymes and increase triglycerides. In non-lipogenic tissues, PUFAs have been reported to affect the *Thy-1* antigen on T-lymphocytes, fatty acid binding proteins, apolipoproteins A-IV and C-III, the sodium channel gene in cardiac myocytes, acetyl-CoA carboxylase in pancreatic cells, and steaoryl-CoA desaturase in brain tissue (Gill and Valivety 1997, Sesler and Ntambi 1998). The stimulation of superoxide dismutase by DHA and other PUFAs (Phylactos et al. 1994) represents a mechanism whereby intracellular antioxidants may be stimulated. This is particularly important in neonatology as infants which normally receive DHA from their mother (prenatally across the placenta, or postnatally via breast milk) have a better antioxidant status than those infants fed a formula without supplemental DHA and ARA. Breast-fed infants appear to be protected from certain oxidant-precipitated pathologies such as necrotising enterocolitis (NEC) (Crawford et al. 1998). This has been correlated with the higher levels of cupric/zinc superoxide dismutase in the erythrocytes of breast-fed vs. formula-fed infants (Phylactos et al. 1995). Consistent with a DHA-stimulation of antioxidant status is the recent observation that feeding infants with a DHA/ARA-supplemented formula results in a remarkable reduction of NEC from an 18% incidence to a mere 3% incidence in one hospital setting (Carlson et al. 1998).

Finally, Leaf and colleagues have suggested that DHA may also play a role in calcium channel regulation (Billman et al. 1997, Xiao et al. 1997). In assays using isolated cardiac myocytes, DHA was shown to be effective in blocking calcium channels which had been opened by ouabain treatment of the cells (Leaf 1995). Billman and colleagues (Billman et al. 1994, 1999) demonstrated that the number of fatal arrhythmias in a dog model of sudden cardiac death could be significantly reduced if the dogs were pre-treated with high doses of DHA. Rat (McLennan et al. 1996) and primate models (Charnock et al. 1992) have also shown similar results. DHA may, therefore, have a secondary role in calcium homeostasis in times of radical neuronal cell injury. Elevated intracellular calcium levels would stimulate a calcium dependant phospholipase which, in turn, would cleave off DHA from the DHA-rich phospholipids in the membranes. The high local concentration of free DHA then closes off the calcium-channel, and the internal calcium levels drop. Although other PUFAs may have similar effects, it is DHA, not EPA, that is enriched in neurological and cardiac cells, and it is therefore more likely that DHA is the active fatty acid in this endogenous control mechanism. Indeed, McLennan and colleagues (McLennan et al. 1996) demonstrated that at low dietary intakes, DHA, not EPA, inhibits ischaemia-induced cardiac arrhythmias in the rat.

3. DHA is an ancient molecule

For the first 2.5 billion years of life on this planet, the atmosphere was highly reducing and contained very little molecular oxygen. Single cell life forms dominated by blue-green algae existed in the proto oceans covering the planet. These algae utilised light energy and reduced carbon and nitrogen sources for the production of more complex molecules including proteins, carbohydrates and lipids. These lipids were rich in PUFAs, particularly of the omega-3 family. The omega-3 family of fatty acids contain the most unsaturated fatty acids used in biological systems. Generally, PUFAs are very sensitive to damage by oxygen radicals, but the absence of oxygen in this ancient environment resulted in little danger of these components being oxidised. The process of photosynthesis itself, however, produced ever-increasing amounts of molecular oxygen with time, and the atmosphere slowly became more oxidising. Due to the increase in atmospheric oxygen, about 600 million years ago there appeared a new and more efficient biochemical mechanism to produce cellular energy – oxidative metabolism. This new efficiency resulted in an explosion of new phyla in the fossil record. Thus, the visual and nervous systems of all species originated in an environment which was rich in very long chain, omega-3 PUFAs, such as DHA.

The dominance of omega-3 fatty acids in the oceans still exists today, and these fatty acids are still a requirement for reproduction and early development of all marine species. Even the early terrestrial plant species (e.g., ferns, mosses) were dependent on water and omega-3 PUFAs for reproduction. About 70 million years ago, new terrestrial plant species appeared with "protected" seeds, eliminating the requirement for abundant amounts of water for reproduction. The seeds of these new, flowering plants, were also better adapted to life in the more oxidising terrestrial environment. Their lipids contained less unsaturated fatty acids and were, thereby, more resistant to oxidative damage. These fatty acids with fewer double bonds were dominated by the omega-6 family. As a result of this change in fatty acid composition of available food sources, the evolving terrestrial animal species had less omega-3 fatty acids and more of the omega-6 fats in their diet. Consequently, in many terrestrial animal species, the omega-6 components, not the omega-3 components became essential for reproduction (Crawford and Marsh 1995). DHA, however, still remained the major component of the nervous system.

The abundance of omega-6 fatty acids and the dearth of omega-3 fatty acids in the nutrition of terrestrial animals resulted in a great increase in the body weight to brain weight ratios (the encephaliation quotient; EQ) (Crawford 1993). The mammalian brain size, however, is generally still larger in relation to body size compared to the egg laying amphibians, reptiles and fish. This enigma may be explained by the evolution of the mammalian placenta. The placenta is an organ that concentrates specific nutrients (*e.g.*, DHA and ARA), and energy, from the mother and transfers these components to her progeny *in utero*, to get them through a critical period of brain development. In other words, the mammalian placenta robs the DHA accumulated by the mother in an environment already poor in DHA, and provides an environment rich in DHA (like the marine environment) for the developing offspring. The emergence of the vascularized placenta, therefore, was required for the development of the larger brains of the mammals in a terrestrial, omega-3 deficient environment.

As the body masses of the evolving terrestrial mammals increased, the EQ dropped off precipitously due to the unavailability of the building blocks of the brain, namely DHA, in the diet. These differences are so great, that the availability of DHA can be considered as a limiting factor in the evolution of the brain. The marine mammals, on the other hand, have a much larger EQ compared to terrestrial mammals of the same weight. This contrary situation is likely due to the great abundance of the DHA in the marine food web relative to the terrestrial food web. Although primates, like other mammals, also follow this rule of the EQ, the single most important and somewhat perplexing exception is hominids. This may point to a novel evolutionary pathway taken by the human primates.

4. The enigma of the modern human brain

The African savannah ecosystem has long been thought to be the origin of the hunter/gatherer progenitor of man. Since man does not have the capacity to synthesise DHA from the abundant omega-6 precursors in the environment, nor is omega-6 docosapentaenoic acid (DPA; the omega-6 version of DHA) used in its place, the expansion of the human brain required a plentiful source of pre-formed DHA in the diet. Crawford and colleagues have therefore proposed that *Homo sapiens* could not have evolved on the savannahs, but rather, the transition from the archaic to modern humans must have taken place at the land/water interface (Broadhurst et al. 1998).

Australopithecus spp. existed for over 3 million years with a relatively small brain (paleontological records indicate that the cranial capacity was only about 500 cm^3). In a span of only about 1 million years, however, the cranial capacity more than doubled to about 1250 cm^3 as seen in *H. sapiens* (Broadhurst et al. 1998). This increase in brain capacity took place with only minor changes in body weight. The sudden increase in the EQ in the last 200 000 years can be explained by a dramatic change in the ecological niche occupied by hominids from one poor in DHA to one rich in DHA. Such an ecological niche could only be the land/water interface. The paleontological record supports this speculation in that the earliest evidence for modern *H. sapiens* found in Africa are associated with lake shore environments in the East African Rift Valley.

The concept of man as an aquatic ape has been based on other physiological enigmas inconsistent with the slow Darwinian evolution of man on the savannahs of Africa. For example, it is counterintuitive that an animal evolving in an arid region should lose water at extraordinarily high rates by perspiration and through dilute urine as is the case for *H. sapiens*. No other animals living in arid regions have such high rates of water and salt loss. The erect posture of man is also of no benefit to hunting success in a savannah environment. Indeed, it may be a significant detriment, as the most successful hunters crouch, hide, and use stealth to capture their prey in such an open environment. Erect posture would, however, emerge as a natural consequence of working in coastal water to obtain food. These and other characteristics are more adaptive to a water environment suggesting that *H. sapiens* may have evolved near water (Broadhurst et al. 1998, Crawford 1993). The association between the rapid increase in brain capacity and the occupation of the land/water interface by early *H. sapiens* reflects of the dramatic influence of brain specific nutrition on the evolutionary process. It is not likely that *H. sapiens* evolved a large, complex brain, and then began to hunt for the nutritional components for its maintenance. Rather, the large brain was more likely to have developed as a consequence of the abundant availability of the building blocks for that structure (*e.g.*, DHA). In this case, brain specific nutrition was a driving force of evolution (Crawford and Marsh 1995).

5. Clinical consequences of hypodocosahexaenemia

In order to help understand the critical importance and biochemical function of DHA in the brain, we have asked the question – what are the physiological manifestations of sub-optimal levels of DHA, or hypodocosahexaenemia? This is a particularly important question as we recognise that the present Western diet contains dramatically less DHA than the diets to which our species evolved. We are no longer living exclusively at the land/water interface, and we are no longer eating the DHA-rich diets that allowed our large-brained hominid ancestors to evolve. Is this a problem?

5.1. Animal models

Animal studies have demonstrated that when using a feeding regimen completely devoid of omega-3 fatty acids, the brain DHA levels are dramatically reduced. It is replaced with the omega-6 counterpart, n-6 DPA. Under these conditions, animals have more difficulty with learning tasks (Fujimoto et al. 1989). The first demonstration of a behavioral disturbance was self-mutilation in Capuchin monkeys (Fiennes et al. 1973). It is impossible to say if the self-mutilation was the specific result of brain deficits, or secondary to other responses. Visual function is also adversely affected by subnormal levels of DHA in the brains and/or the retinas of animals that have been maintained on an omega-3 deficient diet (Neuringer et al. 1988). Even long-term behavioral changes may be associated with deficiencies of DHA early in life during the critical periods of brain development. Higley and co-workers (Higley et al. 1996a,b) demonstrated that infant Rhesus monkeys who are fed their own mother's milk (a dietary supply of DHA) and who are nurtured by their mother for the first three months of life, were less aggressive, less depressed in adolescence, and always achieved a much higher social rank as adults when compared to animals who were raised with a peer group and fed a formula missing DHA. This behavioral change was shown to be related to a depressed level of brain serotonin (measured by cerebral spinal fluid 5-hydroxy indol acetic acid, or 5-HIAA levels) which carried over from infancy to adulthood. This was a remarkable finding since the only differences between the two groups was the feeding regimen and the mother's nurturing for the first 3 months of life. After that time, all monkeys were treated the same. There is recent evidence in support of a similar role for DHA in the reduction of stress-induced aggression in humans (Hamazaki et al. 1996).

The biochemical basis for such long-term effects of an early DHA deficiency insult may be partially explained by the recent work of Heinz and co-workers (Xu et al. 1996). A fatty acid binding protein with a very high specificity for DHA was shown to be expressed at high levels by glial cells in mice early in the development of the brain. Furthermore, if this DHA binding protein was neutralised by an antibody, the neurons did not migrate normally from the ventricles to the cortical plate during brain differentiation. Later in life, this protein is not expressed at high levels except in the case of acute brain injury.

5.2. Clinical manifestations in human brain function

There are many different human neuropathologies that are associated with reduced levels of DHA. Children diagnosed with attention deficit hyperactivity disorder (ADHD) have subnormal serum levels of DHA and ARA (Burgess et al. 2000). Furthermore, the lower the blood levels of DHA in these children, the more prevalent were the hyperactivity symptoms. The appearance of ADHD in epidemic proportions in the US in the last twenty years coincides with the increasing usage of infant formulas containing no supplemental DHA. Indeed, Burgess and co-workers further reported that the ADHD group of children were much more likely to have been formula-fed than breast-fed, and the control group were much more likely to have been breast-fed than formula-fed (Stevens et al. 1995). McCreadie (McCreadie 1997) also recently showed that formula feeding (lack of early DHA supplementation) is also a significant risk factor for the later development of schizophrenia or schizoid tendencies. Like the infant rhesus monkey data of Higley (Higley and Linnoila 1997), a DHA deficiency at an early, but critical stage of neurodevelopment, may have significant long-term behavioral consequences.

There is also a correlation between depression and low levels of circulating DHA. A possible mechanism for this relationship was suggested by Hibbeln, who reported a correlation between plasma DHA content and levels of 5-HIAA in the cerebral spinal fluid of healthy adults (Hibbeln et al. 1998a). That is, the higher the serum DHA levels, the higher were the serotonin levels as measured by CSF 5-HIAA. Interestingly, Hibbeln also showed that the opposite was true for violently aggressive patients (Hibbeln et al. 1998b). Although not a controlled intervention, a world-wide, cross cultural comparison of the incidence of major depressive episodes and dietary fish intake (this reflects dietary DHA intake as fish represents the primary source of DHA) indicated that there is a highly significant ($p < 0.0001$) negative correlation between fish (DHA) consumption and frequency of major depression (Hibbeln 1998). Once again, the greater the DHA intake, the lower was the frequency of major lifetime depressive episodes. Hibbeln also showed a similar correlation between fish (DHA) in the diet and postpartum depression in women.

Controlled dietary intervention studies are presently underway and preliminary data appear to support the epidemiological findings described above. A supplementation study in patients with bipolar disorder using oils rich in both DHA and EPA resulted in a remarkable improvement, and significant delay in the recurrence of symptoms (Stoll et al. 1999). In an unrelated study, women were given DHA supplements immediately after birth, and a significant improvement in Stroop assessments of concentration were observed after 12 weeks compared to a placebo-treated control group (Jensen et al. 1999). Women in the latter study that had a Beck Depression Inventory (BDI) assessment of greater than 10 at 2 weeks postpartum, also exhibited a significant reduction in the BDI by 8 weeks in a post-hoc analysis, compared to a placebo treated control group.

Several other neurological pathologies have been reported to be associated with hypodocosahexaenemia. These include schizophrenia (Glen et al. 1994, Laugharne et al. 1996), senile dementia (Kyle et al. 1999), and tardive dyskinesia (Vaddadi et al. 1989). In the latter case, Vaddadi and co-workers demonstrated that the symptomology of tardive dyskinesia was most severe in individuals with the lowest DHA levels and that supplementation with omega-3 PUFAs, including DHA, significantly improved this condition. The brain tissues of patients who were diagnosed with Alzheimer's Dementia (AD) were found to contain about 30% less DHA (especially the hippocampus and frontal lobes) compared to similar tissue isolated from pair-matched geriatric controls (Prasad et al. 1998, Soderberg et al. 1991). In a 10-year prospective study with about 1200 elderly patients monitored regularly for signs of the onset of dementia, a low serum phosphatidylcholine DHA (PC–DHA) level was a significant risk factor for the onset of senile dementia (Kyle et al. 1999). Individuals with plasma PC–DHA levels less than 3.5% of total fatty acids (*i.e.,* the lower half of the DHA distribution), had a 67% greater risk of being diagnosed with senile dementia in the subsequent 10 years compared to those with DHA levels higher than 3.5%. Furthermore, for women who had at least one copy of the Apolipoprotein E4 allele, the risk of getting a low score on the minimental state exam (MMSE) went up by 400% if their plasma PC–DHA levels were also less than 3.5%.

Finally, certain types of peroxisomal disorders may represent the most profound examples of DHA deficiency found in neurology (Martinez 1991, 1992). Since the last step of DHA biosynthesis requires peroxisomal function, diseases such as Zellweger's, Refsums's, and neonatal adrenoleukodystrophy which are characterised by peroxisomal dysfunction, are accompanied by a severe DHA deficiency. Preliminary findings of Martinez and colleagues indicate that the development of some of these progressive neurodegenerative diseases can be arrested if DHA is reintroduced into the diet at an early stage (Martinez et al. 1993). Much of the neurodegeneration is manifest in a progressive demyelination which, until now, was thought to be irreversible. Martinez and co-workers (Martinez and Vazquez 1998) have recently reported that not only do symptoms improve upon treatment of certain of these patients with DHA, but magnetic resonance imaging data indicate that the brain begins to re-myelinate once DHA therapy is initiated.

5.3. Clinical manifestations in human visual function

Visual tissues are an extension of the tissues of the central nervous system and there are several visual pathologies which are also associated with abnormally low levels of circulating DHA. These include retinitis pigmentosa (Hoffman et al. 1995), long chain hydroxyacyl-CoA dehydrogenase deficiency (LCHADD) (Harding et al. 1999), macular degeneration, and even dyslexia (Stordy 2000). Individuals with dyslexia generally also have a poor night vision. In one DHA supplementation study with dyslexic adults, significant improvements in dark adaptation (night vision) in a DHA (and EPA) supplemented group were demonstrated compared to an unsupplemented control group (Stordy 1995). More recently, Gillingham has reported results from a supplementation study using DHA alone in patients with LCHADD (Gillingham et al. 1999). This disease results in a very low level of circulating DHA in the plasma, and these patients progressively go blind. Supplementation with DHA not only arrested the visual decline in 90% of these patients, but it has actually resulted in significant improvements in their visual function. A similar improvement in visual indices has been reported by Martinez in patients suffering visual losses associated with peroxisomal disorders (Martinez 1996). Outcomes in patients with retinitis pigmentosa and macular degeneration are the subjects of ongoing clinical intervention trials.

6. DHA and neurodevelopment in infants

Perhaps at no time in our life history is adequate DHA nutrition more critical than during first few years of life. The massive amount of brain development that takes place in the growing foetus during pregnancy and by the infant over the first two postnatal years requires special dietary delivery mechanisms for DHA and ARA. *In utero*, the human placenta nearly doubles the proportions of DHA and ARA from the maternal blood stream to pass them on to the growing foetus. Specific DHA transport mechanisms pumping DHA across the placenta underscores the developmental need for this component by the foetus. During the period of gestation, the DHA status of the mother declines as she draws on her internal pools of DHA to provide this essential component to the growing foetus (Al et al. 1995). Once the baby is born, the mother continues to draw down her reserves of DHA providing the infant with DHA through her breast milk.

Infants who are fed infant formulas with no supplemental DHA or ARA have dramatically altered blood and brain chemistries compared to infants who are fed breast milk which provides a natural supplement of both DHA and ARA (Table 1). The blood of a formula-fed infant will contain less than half of the DHA of the blood of a breast-fed baby (Carlson et al. 1992, Uauy et al. 1992). Brain DHA levels of formula-fed babies can be as much as 30% lower that those of breast-fed babies (Byard et al. 1995, Farquharson et al. 1995). Thus, feeding an infant a formula without supplemental DHA and ARA as the sole source of nutrition puts that infant into a deficiency state relative to a breast-fed baby. This may be especially problematic for the pre-term infant whose brain is still developing and who is no longer receiving DHA in high concentrations from the mother across the placenta (Crawford 2000, Crawford et al. 1998). Since the brain lipid binding protein responsible for proper neuronal development and migration has a specific requirement for DHA (Xu et al. 1996), an inadequate supply of DHA to the growing foetus *in utero,* and to the infant postnatally, could affect the normal migration patterns during this crucial period of development.

Many studies have attempted to assess the consequences of such a DHA deficiency in the blood and brain of formula-fed infants by comparing the long-term mental outcomes of breast-fed vs. formula-fed babies (Florey et al. 1995, Golding et al. 1997, Horwood and Fergusson 1998, Lanting et al. 1994, Lucas et al. 1992). A meta-analysis of these studies has clearly established that even after all the confounding data, including socio-economic status, sex, birth order, *etc.*, have been taken into account, there is still a small, but significant advantage (3–4 IQ points) for the breast-fed babies when measured by standard IQ assessment, general performance tests (Anderson et al. 1999) or long-term neurological complications (Lanting et al. 1994).

A large number of clinical studies have now also compared biochemical, neurological and visual outcomes of both term and preterm babies fed either formulas containing supplemental DHA (and ARA), or standard infant formulas. Although these have all been relatively small studies (up to a few hundred babies per study), in every single case the standard formula-fed infants exhibited an altered blood chemistry reflected in significantly low DHA and ARA levels than breast-fed or DHA/ARA-supplemented formula-fed babies (Table 1). Furthermore, in most cases developmental delays in visual, or mental acuity were also detected with the standard formula-fed babies (Table 1). Where these differences were observed, they were always overcome in babies who were fed DHA/ARA-supplemented formulas. In the most recent of these studies, a 7 point improvement in the Baily Mental Development Index (MDI) was seen in infants fed formulas supplemented with DHA and ARA over those infants receiving standard formula (Birch et al. 2000). It is interesting to note, however, that the DHA-fortified formula-fed babies were never better off than the breast-fed babies were. Rather, it was the deficit caused by the standard formula feeding that was overcome by the supplementation of the formula with DHA and ARA.

Although the majority of the studies listed in Table 1 underscore the problem associated with standard formula feeding, there were some exceptions of note. In one study (Auestad et al. 1997), the level of DHA supplementation used in their DHA-supplemented formula group was only 0.12%. This level represents the lowest level in the range of breast milk DHA levels (0.1%–0.9%) which the authors quote, and far less than the supplementation levels recommended by an Expert Committee of the FAO/WHO[1]. Most of the other intervention studies shown in Table 1 used DHA and ARA doses that were more physiological. The second study (Horby Jorgensen et al. 1998) did report a trend toward improvement in the visual outcomes of babies fed a DHA-supplemented formula over the standard formula, but the sample size was too small for it to attain statistical significance.

Recent studies by Lucas (Lucas et al. 1999) and Makrides (Makrides et al. 2000) have begun to clarify the importance of the complex lipid form of delivery of DHA and ARA to the infant. DHA and ARA are naturally delivered to the infant in breast milk as a triglyceride. The Lucas study provided DHA and ARA in a phospholipid form (egg yolk) and they reported no differences in Bailey MDI between supplemented and unsupplemented infants. Makrides compared DHA delivered as a triglyceride to DHA delivered as a phospholipid in the same study and found a 6 point higher MDI score in the infants receiving the triglyceride form compared to the egg yolk. In fact, there was no significant difference in MDI scores between breast-fed babies and those receiving the triglyceride form of DHA, whereas the babies receiving the phospholipid form of

---

[1] Report #57 of a Joint Expert Consultation of the Food and Agriculture Organization and the World Health Organization (Rome, October 1993) entitled Fats and Oils in Human Nutrition.

Table 1

A comprehensive list of clinical intervention trials using infant formulas supplemented with DHA and ARA in comparison to standard formulas and breast-fed babies[a]

Study[b]	PUFA source	Size (n)	Are babies' blood lipids normalized with DHA/AA?		Are functional outcomes normalized with DHA/AA?	
			DHA	AA	Visual	Neurological
Preterm						
(1)	egg	29	Yes	Yes	not tested	not tested
(2)	fish	32	Yes	Yes	not tested	not tested
(3)	fish	51	Yes	Yes	Yes	not tested
(4)	SCO	16	Yes	Yes	not tested	not tested
(5)	SCO	43	Yes	Yes	not tested	not tested
(6)	egg	41	Yes	Yes	not tested	not tested
(7)	fish	59	Yes	Yes	Yes	not tested
(8)	SCO	284	Yes	Yes	no difference	not tested
(9)	SCO	91	Yes	Yes	not tested	not tested
(10)	SCO	287	Yes	Yes	not tested	not tested
(11)	egg	119	Yes	Yes	not tested	not tested
Term						
(12)	egg	n.g.	Yes	Yes	not tested	not tested
(13)	egg	90	Yes	Yes	not tested	Yes
(14)	egg	22	Yes	Yes	not tested	not tested
(15)	fish	55	Yes	Yes	Yes	not tested
(16)	egg	58	Yes	Yes	Yes	not tested
(17)	fish	131	Yes	Yes	not tested	not tested
(18)	SCO	113	Yes	Yes	not tested	not tested
(19)	egg	197	Yes	Yes	no difference	no difference
(20)	fish	67	Yes	Yes	no difference	not tested
(21)	egg	123	Yes	Yes	not tested	not tested
(22)	egg	44	not given	not given	not tested	Yes
(23)	SCO	108	Yes	Yes	not tested	Yes
(24)	fish	56	Yes	Yes	No difference	not tested

[a] In all studies the babies' blood DHA and ARA levels were significantly lower with standard formula feeding compared to breast feeding and in many cases the neurological and visual functional outcomes were also poorer. The table addresses the questions of whether DHA/ARA supplemented formulas can return the babies' blood lipids or functional outcomes to normal as defined by the breast-fed babies (the gold standard). Sources of DHA and ARA supplementation for the formulas include fish oil, egg yolk lipid and single cell oil (SCO). (n.g.) data not given.

[b] References: *Preterm Infants*: 1. Koletzko et al. (1989) 2. Clandinin et al. (1992) 3. Hoffman et al. (1993) 4. Carnielli et al. (1994) 5. Foreman-van Drongelen et al. (1996) 6. Boehm et al. (1996) 7. Carlson et al. (1996) 8. Hansen et al. (1997) 9. Clandinin et al. (1997) 10. Vanderhoof et al. (1999) 11. Carlson et al. (1998).
Term Infants: 12. Kohn et al. (1994) 13. Agostoni et al. (1995) 14. Decsi and Koletzko (1995) 15. Makrides et al. (1995) 16. Carlson and Werkman (1996) 17. Innis and Hansen (1996) 18. Gibson et al. (1997b) 19. Auestad et al. (1997) 20. Gibson et al. (1997a) 21. Bellu (1997) 22. Willatts et al. (1998) 23. Birch et al. (1998) 24. Horby Jorgensen et al. (1998).

DHA scored significantly lower than the breast-fed babies and were not different from the unsupplemented formula-fed babies.

Many of the studies shown in Table 1 also reported that the developmental delays eventually "caught-up" after some time. However, it is not clear whether there was a true normalisation of the function, or if the tests were simply no longer sensitive for the stage of development for which they were being used. Thus, unsupplemented formula-fed babies may be at a significant disadvantage because of the DHA deficiency in the brain and eyes during this early period of life.

As the result of the need for supplemental DHA and ARA by infants who are not receiving their mother's milk, several organisations have made recommendations that all infant formulas contain supplemental DHA and ARA at the levels normally found in mother's milk. Of particular importance was the recommendation by a joint select committee of the Food and Agriculture Organisation and the World Health Organisation which drew on the expertise of nearly fifty researchers in this field, and concluded that adequate dietary DHA was also important for the mother postnatally, prenatally and even preconceptually. More recently, another consensus conference recommended that to meet the requirements of a growing infant, infant formulas should contain 0.35% of lipid as DHA and 0.5% of lipid as ARA (Simopoulos et al. 1999). Both of these bodies also pointed out the importance of maintaining an adequate DHA supply to the mother during pregnancy and the latter groups suggested that this could be met with 300 mg DHA/day. The clinical data and subsequent recommendations have resulted in the replacement of many of the older standard infant formulas around the world with new formulas supplemented with DHA and ARA at the same levels as found in breast milk.

7. Conclusions

DHA is a molecule that is found universally in neural and visual tissues in all animals so far studied. We suggest that this is no accident of biology, but rather the result of specific biochemical and physical attributes of this highly unusual fatty acid. It likely had its origin in the single cell life inhabiting the primordial proto-oceans when the atmosphere was primarily reducing. Later, it remained as the fundamental building block of all tissues with electrical activity. Its availability in sea and lacustrine foods may have been a key factor in the rapid expansion of the brain capacity of early *Homo* species.

As a result of its functional importance, and as a consequence of man's limited ability to synthesise DHA *de novo*, we have become dependant on the availability of dietary DHA for the optimal development of the brain. Several disorders of the neuro/visual axis have been correlated with low levels of tissue DHA and intervention trials have demonstrated that repletion of the diet with DHA can ameliorate these conditions. Early data in the area of depression, or mood disorders, and in certain visual disorders, suggest that dietary repletion can significantly improve the morbidity associated with these conditions.

Perhaps the most important issue today concerning DHA and brain development, however, relates to the appropriate feeding of infants. Breast milk represents the optimal nutritional source for a growing infant and it provides DHA and ARA for the development of the baby's brain. Until recently, artificial infant formulas did not contain any supplemental DHA. The neurological outcomes of formula-fed babies has consistently been found to be poorer than that of breast-fed babies. New data on the importance of specific DHA binding proteins for proper neuronal migration during brain development, and animal studies describing profound long-term differences in neurotransmitter levels as a consequence of early nutrition, are providing biochemical explanations for these early observations. As a result of these new data and recommendations by expert panels, standard infant formulas are being replaced by ARA and DHA-supplemented formulas, thereby providing formula-fed babies with a nutritional composition closer to mother's milk and one that provides better nutrition for the brain.

Acknowledgements

The author wishes to acknowledge the contributions, conversations, and critical review of Professor Michael Crawford whose thoughts guided much of this manuscript, and to the editorial review of Dr. Linda Arterburn.

References

Agostoni, C., Trojan, S., Bellu, R., Riva, E. and Giovannini, M. (1995) Neurodevelopmental quotient of healthy term infants at 4 months and feeding practice: the role of long-chain polyunsaturated fatty acids. Pediatr. Res. 38, 262–266.

Agren, J.J., Hanninen, O., Julkunen, A., Fogelholm, L., Vidgren, H., Schwab, U., Pynnonen, O. and Uusitupa, M. (1996) Fish diet, fish oil and docosahexaenoic acid rich oil lower fasting and postprandial plasma lipid levels. Eur. J. Clin. Nutr. 50, 765–771.

Al, M.D., van Houwelingen, A.C., Kester, A.D., Hasaart, T.H., de Jong, A.E. and Hornstra, G. (1995) Maternal essential fatty acid patterns during normal pregnancy and their relationship to the neonatal essential fatty acid status. Br. J. Nutr. 74, 55–68.

Anderson, J.W., Johnstone, B.M. and Remley, D.T. (1999) Breast-feeding and cognitive development: a meta-analysis. Am. J. Clin. Nutr. 70, 525–535.

Arbuckle, L.D. and Innis, S.M. (1993) Docosahexaenoic acid is transferred through maternal diet to milk and to tissues of natural milk-fed piglets. J. Nutr. 123, 1668–1675.

Auestad, N., Montalto, M.B., Hall, R.T., Fitzgerald, K.M., Wheeler, R.E., Connor, W.E., Neuringer, M., Connor, S.L., Taylor, J.A. and Hartmann, E.E. (1997) Visual acuity, erythrocyte fatty acid composition, and growth in term infants fed formulas with long chain polyunsaturated fatty acids for one year. Ross Pediatric Lipid Study. Pediatr. Res. 41, 1–10.

Bazan, N.G. and Rodriguez de Turco, E.B. (1994) Review: pharmacological manipulation of docosahexaenoic-phospholipid biosynthesis in photoreceptor cells: implications in retinal degeneration. J. Ocul. Pharmacol. 10, 591–604.

Bazan, N.G. and Scott, B.L. (1990) Dietary omega-3 fatty acids and accumulation of docosahexaenoic acid in rod photoreceptor cells of the retina and at synapses. Ups. J. Med. Sci. Suppl. 48, 97–107.

Bazan, N.G., Rodriguez de Turco, E.B. and Gordon, W.C. (1994) Docosahexaenoic acid supply to the retina and its conservation in photoreceptor cells by active retinal pigment epithelium-mediated recycling. World Rev. Nutr. Diet. 75, 120–123.

Behrens, P. and Kyle, D. (1996) Microalgae as a source of fatty acids. J. Food Sci. 3, 259–272.

Bellu, R. (1997) Effects of a formula supplemented with LC-PUFA on growth and anthropometric indices in infants from 0 to 4 months: a prospective randomized trial. J. Am. Coll. Nutr. 16, 490.

Billman, G.E., Hallaq, H. and Leaf, A. (1994) Prevention of ischemia-induced ventricular fibrillation by omega 3 fatty acids. Proc. Natl. Acad. Sci. U.S.A. 91, 4427–4430.

Billman, G.E., Kang, J.X. and Leaf, A. (1997) Prevention of ischemia-induced cardiac sudden death by n-3 polyunsaturated fatty acids in dogs. Lipids 32, 1161–1168.

Billman, G.E., Kang, J.X. and Leaf, A. (1999) Prevention of sudden cardiac death by dietary pure omega-3 polyunsaturated fatty acids in dogs. Circulation 99, 2452–2457.

Birch, E., Hoffman, D., Uauy, R., Birch, D. and Prestidge, C. (1998) Visual acuity and the essentiality of docosahexaenoic acid and arachidonic acid in the diet of term infants. Pediatr. Res. 44, 201–209.

Birch, E.E., Garfield, S., Hoffman, D.R., Uauy, R. and Birch, D.G. (2000) A randomized controlled trial of early dietary supply of long chain polyunsaturated fatty acids and mental development in term infants. Hum. Dev. Child Neurol. 70, 174–181.

Bloom, M., Linseisen, F., Lloyd-Smith, J. and Crawford, M. (1999) Insights from NMR on the functional role of poluunsaturated lipids in the brain. In: B. Maravigla (Ed.), Magnetic Resonance and Brain Function – Approaches from Physics. Proc. 1998 Enrico Fermi Int. School of Physics, Enrico Fermi Lecture, Course 139, Varenna, Italy, IOS Press, Italy, pp. 527–553.

Boehm, G., Borte, M., Bohles, H.J., Muller, H. and Moro, G. (1996) Docosahexaenoic and arachidonic acid content of serum and red blood cell membrane phospholipids of preterm infants fed breast milk, standard formula or formula supplemented with n-3 and n-6 long-chain polyunsaturated fatty acids. Eur. J. Pediatr. 155, 410–416.

Broadhurst, C.L., Cunnane, S.C. and Crawford, M.A. (1998) Rift Valley lake fish and shellfish provided brain-specific nutrition for early Homo. Br. J. Nutr. 79, 3–21.

Burgess, J.R., Stevens, L., Zhang, W. and Peck, L. (2000) Long-chain polyunsaturated fatty acids in children with attention-deficit hyperactivity disorder. Am. J. Clin. Nutr. 71, 327S–330S.

Byard, R.W., Makrides, M., Need, M., Neumann, M.A. and Gibson, R.A. (1995) Sudden infant death syndrome: effect of breast and formula feeding on frontal cortex and brainstem lipid composition. J. Paediatr. Child Health 31, 14–16.

Carlson, S.E. and Werkman, S.H. (1996) A randomized trial of visual attention of preterm infants fed docosahexaenoic acid until two months. Lipids 31, 85–90.

Carlson, S.E., Cooke, R.J., Rhodes, P.G., Peeples, J.M. and Werkman, S.H. (1992) Effect of vegetable and marine oils in preterm infant formulas on blood arachidonic and docosahexaenoic acids. J. Pediatr. 120, S159–S167.

Carlson, S.E., Werkman, S.H., Peeples, J.M. and Wilson, W.M. (1994) Long-chain fatty acids and early visual and cognitive development of preterm infants. Eur. J. Clin. Nutr. 48, S27–S30.

Carlson, S.E., Ford, A.J., Werkman, S.H., Peeples, J.M. and Koo, W.W. (1996) Visual acuity and fatty acid status of term infants fed human milk and formulas with and without docosahexaenoate and arachidonate from egg yolk lecithin. Pediatr. Res. 39, 882–888.

Carlson, S.E., Montalto, M.B., Ponder, D.L., Werkman, S.H. and Korones, S.B. (1998) Lower incidence of necrotizing enterocolitis in infants fed a preterm formula with egg phospholipids. Pediatr. Res. 44, 491–498.

Carnielli, V.P., Tederzini, F., Luijendijk, I.H.T., Bomaars, W.E.M., Boerlage, A., Degenhart, H.J., Pedrotti, D. and Sauer, P.J.J. (1994) Long chain polyunsaturated fatty acids (LCP) in low birth weight formula at levels found in human colostrum. Pediatr. Res. 35, 309A.

Charnock, J.S., McLennan, P.L. and Abeywardena, M.Y. (1992) Dietary modulation of lipid metabolism and mechanical performance of the heart. Mol. Cell Biochem. 116, 19–25.

Clandinin, M.T., Garg, M.L., Parrott, A., Van Aerde, J., Hervada, A. and Lien, E. (1992) Addition of long-chain polunsaturated fatty acids to formula for very low birth weight infants. Lipids 27, 869–900.

Clandinin, M.T., Van Aerde, J.E., Parrott, A., Field, C.J., Euler, A.R. and Lien, E.L. (1997) Assessment of the efficacious dose of arachidonic and docosahexaenoic acids in preterm infant formulas: fatty acid composition of erythrocyte membrane lipids. Pediatr. Res. 42, 819–825.

Clarke, S. and Jump, D. (1994) Dietary polyunsaturated fatty acid regulation of gene transcription. Annu. Rev. Nutr. 14, 83–89.

Connor, W.E., Weleber, R.G., Lin, D.S., Defrancesco, C. and Wolf, D.P. (1995) Sperm abnormalities in retinitis pigmentosa. J. Invest. Med. 43, 302A.

Conquer, J.A. and Holub, B.J. (1996) Supplementation with an algae source of docosahexaenoic acid increases (n-3) fatty acid status and alters selected risk factors for heart disease in vegetarian subjects. J. Nutr. 126, 3032–3039.

Conquer, J.A., Martin, J.B., Tummon, I., Watson, L. and Tekpetey, F. (1999) Fatty acid analysis of blood serum, seminal plasma, and spermatozoa of normozoospermic vs. asthenozoospermic males. Lipids 34, 793–799.

Crawford, M.A. (1993) The role of essential fatty acids in neural development: implications for perinatal nutrition. Am. J. Clin. Nutr. 57, 703S–709S.

Crawford, M.A. (2000) Placental delivery of arachidonic and docosahexaenoic acids: implications for the lipid nutrition of preterm infants. Am. J. Clin. Nutr. 71, 275S–284S.

Crawford, M.A. and Marsh, D. (1995) Nutrition and Evolution, Keats Publishing, New Canaan, USA.

Crawford, M.A., Costeloe, K., Ghebremeskel, K. and Phylactos, A. (1998) The inadequacy of the essential fatty acid content of present preterm feeds. Eur. J. Pediatr. 157(Suppl. 1), S23–S27. Erratum: 157(2), 160.

Cunnane, S.C., Francescutti, V. and Brenna, J.T. (1999) Docosahexaenoate requirement and infant development [letter]. Nutrition 15, 801–802.

Davidson, M.H., Maki, K.C., Kalkowski, J., Schaefer, E.J., Torri, S.A. and Drennan, K.B. (1997) Effects of docosahexaenoic acid on serum lipoproteins in patients with combined hyperlipidemia: a randomized, double-blind, placebo-controlled trial. J. Am. Coll. Nutr. 16, 236–243.

Decsi, T. and Koletzko, B. (1995) Growth, fatty acid composition of plasma lipid classes, and alpha-tocopherol concentrations in full-term infants fed formula enriched with omega-6 and omega-3 long-chain polyunsaturated fatty acids. Acta Paediatr. 84, 725–732.

Farquharson, J., Jamieson, E.C., Abbasi, K.A., Patrick, W.J., Logan, R.W. and Cockburn, F. (1995) Effect of diet on the fatty acid composition of the major phospholipids of infant cerebral cortex. Arch. Dis. Child. 72, 198–203.

Fiennes, R., Sinclair, A. and Crawford, M. (1973) Essential fatty acid studies in primates. Linolenic acid requirements of Capuchins. J. Med. Primatol. 2, 155–169.

Florey, C.D., Leech, A.M. and Blackhall, A. (1995) Infant feeding and mental and motor development at 18 months of age in first born singletons. Int. J. Epidemiol. 24, S21–S26.

Foreman-van Drongelen, M.M., van Houwelingen, A.C., Kester, A.D., Blanco, C.E., Hasaart, T.H. and Hornstra, G. (1996) Influence of feeding artificial-formula milks containing docosahexaenoic and arachidonic acids on the post-natal long-chain polyunsaturated fatty acid status of healthy preterm infants. Br. J. Nutr. 76, 649–667.

Fujimoto, K., Yao, K., Miyazawa, T., Hirono, H., Nishikawa, M., Kimura, S., Maruyama, K. and Nonaka, M. (1989) The Effect of Dietary Docosahexaenoate on the Learning Ability of Rats, Vol. I: Health Effects of Fish and Fish Oils, ARTS Biomedical Publishers and Distributors, St. John's, Newfoundland, Canada.

Gibson, R., Makrides, M., Neuman, M., et al. (1997a) A dose response study of arachidonic acid in formulas containing docosahexaenoic acid in term infants. Prostaglandins Leukotrienes Essent. Fatty Acids 57, 198A.

Gibson, R., Neumann, M. and Makrides, M. (1997b) A randomized clinical trial of LC-PUFA supplements in term infants: effect on neural indices. 88th American Oil Chemists' Society, Seattle, Washington, p. 5.

Gill, I. and Valivety, R. (1997) Polyunsaturated fatty acids, Part 1: Occurrence, biological activities and applications. Trends Biotechnol. 15, 401–409.

Gillingham, M., Van Calcar, S., Ney, D., Wolff, J. and Harding, C. (1999) Dietary management of long-chain 3-hydroxyacyl-CoA dehydrogenase deficiency (LCHADD). A case report and survey. J. Inherit. Metab. Dis. 22, 123–131.

Glen, A.I.M., Glen, E.M.T., Horrobin, D.F., Vaddadi, K.S., Spellman, M., Morse-Fisher, N. and Ellis, K. (1994) A red cell membrane abnormality in a subgroup of schizophrenic patients: Evidence for two diseases. Schizophrenia Res. 12, 53–61.

Golding, J., Rogers, I.S. and Emmett, P.M. (1997) Association between breast feeding, child development and behaviour. Early Hum. Dev. 29, S175–S184.

Gordon, W.C. and Bazan, N.G. (1990) Docosahexaenoic acid utilization during rod photoreceptor cell renewal. J. Neurosci. 10, 2190–2202.

Green, P., Kamensky, B. and Yavin, E. (1997) Replenishment of docosahexaenoic acid in n-3 fatty acid-deficient fetal rats by intraamniotic ethyl-docosahexaenoate administration. J. Neurosci. Res. 48, 264–272.

Greiner, R.C., Winter, J., Nathanielsz, P.W. and Brenna, J.T. (1997) Brain docosahexaenoate accretion in fetal baboons: bioequivalence of dietary alpha-linolenic and docosahexaenoic acids. Pediatr. Res. 42, 826–834.

Greiner, R.S., Moriguchi, T., Hutton, A., Slotnick, B.M. and Salem Jr, N. (1999) Rats with low levels of brain docosahexaenoic acid show impaired performance in olfactory-based and spatial learning tasks. Lipids 34, S239–S243.

Gudbjarnason, S., Doell, B. and Oskarsdottir, G. (1978) Docosahexaenoic acid in cardiac metabolism and function. Acta Biol. Med. Germ. 37, 777–787.

Hamazaki, T., Sawazaki, S., Itomura, M., Asaoka, E., Nagao, Y., Nishimura, N., Yazawa, K., Kuwamori, T. and Kobayashi, M. (1996) The effect of docosahexaenoic acid on aggression in young adults. A placebo-controlled double-blind study. J. Clin. Invest. 97, 1129–1133.

Hansen, J., Schade, D., Harris, C., et al. (1997) Docosahexaenoic acid plus arachidonic acid enhance preterm infant growth. Prostaglandins Leukotrienes Essent. Fatty Acids 57, 196A.

Harding, C.O., Gillingham, M.B., van Calcar, S.C., Wolff, J.A., Verhoeve, J.N. and Mills, M.D. (1999) Docosa-hexaenoic acid and retinal function in children with long-chain 3-hydroxyacyl-CoA dehydrogenase deficiency. J. Inherit. Metab. Dis. 22, 276–280.

Hibbeln, J.R. (1998) Fish consumption and major depression [letter]. Lancet 351, 1213.

Hibbeln, J.R., Linnoila, M., Umhau, J.C., Rawlings, R., George, D.T. and Salem Jr, N. (1998a) Essential fatty acids predict metabolites of serotonin and dopamine in cerebrospinal fluid among healthy control subjects, and early- and late-onset alcoholics. Biol. Psychiatry 44, 235–242.

Hibbeln, J.R., Umhau, J.C., Linnoila, M., George, D.T., Ragan, P.W., Shoaf, S.E., Vaughan, M.R., Rawlings, R. and Salem Jr, N. (1998b) A replication study of violent and nonviolent subjects: cerebrospinal fluid metabolites of serotonin and dopamine are predicted by plasma essential fatty acids. Biol. Psychiatry 44, 243–249.

Higley, J.D. and Linnoila, M. (1997) A nonhuman primate model of excessive alcohol intake. Personality and neurobiological parallels of type I- and type II-like alcoholism. Recent Dev. Alcohol 13, 191–219.

Higley, J.D., Suomi, S.J. and Linnoila, M. (1996a) A nonhuman primate model of type II alcoholism? Part 2. Diminished social competence and excessive aggression correlates with low cerebrospinal fluid 5-hydroxy-indoleacetic acid concentrations. Alcohol Clin. Exp. Res. 20, 643–650.

Higley, J.D., Suomi, S.J. and Linnoila, M. (1996b) A nonhuman primate model of type II excessive alcohol consumption? Part 1. Low cerebrospinal fluid 5-hydroxyindoleacetic acid concentrations and diminished social competence correlate with excessive alcohol consumption. Alcohol Clin. Exp. Res. 20, 629–642.

Hoffman, D.R., Birch, E.E., Birch, D.G. and Uauy, R.D. (1993) Effects of supplementation with omega 3 long-chain polyunsaturated fatty acids on retinol and cortical development in premature infants. Am. J. Clin. Nutr. 57, 807S–812S.

Hoffman, D.R., Uauy, R. and Birch, D.G. (1995) Metabolism of omega-3 fatty acids in patients with autosomal dominant retinitis pigmentosa. Exp. Eye Res. 60, 279–289.

Horby Jorgensen, M., Holmer, G., Lund, P., Hernell, O. and Michaelsen, K.F. (1998) Effect of formula supplemented with docosahexaenoic acid and gamma-linolenic acid on fatty acid status and visual acuity in term infants. J. Pediatr. Gastroenterol. Nutr. 26, 412–421.

Horwood, L.J. and Fergusson, D.M. (1998) Breastfeeding and later cognitive and academic outcomes. Pediatr. 101, E9.

Ikemoto, A., Kobayashi, T., Watanabe, S. and Okuyama, H. (1997) Membrane fatty acid modifications of PC12 cells by arachidonate or docosahexaenoate affect neurite outgrowth but not norepinephrine release. Neurochem. Res. 22, 671–678.

Innis, S.M. and Hansen, J.W. (1996) Plasma fatty acid responses, metabolic effects, and safety of microalgal and fungal oils rich in arachidonic and docosahexaenoic acids in healthy adults. Am. J. Clin. Nutr. 64, 159–167.

Jensen, C., Llorente, A., Voigt, R., et al. (1999) Effects of maternal docosahexaenoic acid (DHA) supplementation on visual and neurodevelopmental function of breast-fed infants and indices of maternal depression and cognitive interference. Pediatr. Res. 39, 284A.

Kohn, G., Sawatzki, G., vanBiervliet, J.P. and Rosseneu, M. (1994) Diet and the essential fatty acid status of term infants. Acta Paediatr. Suppl. 402, 69–74.

Koletzko, B., Schmidt, E., Bremer, H., Huang, M. and Harzer, G. (1989) Effects of dietary long-chain polyunsaturated fatty acid status of premature infants. Eur. J. Pediatr. 147, 669–675.

Kyle, D.J., Schaefer, E., Patton, G. and Beiser, A. (1999) Low serum docosahexaenoic acid is a significant risk factor for Alzheimer's dementia. Lipids 34, S245–S246.

Lamptey, M. and Walker, B. (1976) A possible essential role for dietary linolenic acid in the development of young rats. J. Nutr. 106, 86–93.

Lanting, C.I., Fidler, V., Huisman, M., Touwen, B.C. and Boersma, E.R. (1994) Neurological differences between 9-year-old children fed breast-milk or formula-milk as babies. Lancet 344, 1319–1322.

Laugharne, J.D., Mellor, J.E. and Peet, M. (1996) Fatty acids and schizophrenia. Lipids 31, S163–S165.

Leaf, A. (1995) Omega-3 fatty acids and prevention of ventricular fibrillation. Prostaglandins Leukotrienes Essent. Fatty Acids 52, 197–198.

Lucas, A., Morley, R., Cole, T.J., Lister, G. and Leeson-Payne, C. (1992) Breast milk and subsequent intelligence quotient in children born preterm. Lancet 339, 261–264.

Lucas, A., Stafford, M., Morley, R., Abbott, R., Stephenson, T., MacFadyen, U., Elias-Jones, A. and Clements, H. (1999) Efficacy and safety of long-chain polyunsaturated fatty acid supplementation of infant-formula milk: a randomised trial. Lancet 354, 1948–1954.

Makrides, M., Neumann, M.A., Simmer, K. and Gibson, R.A. (1995) Erythrocyte fatty acids of term infants fed either breast milk, standard formula, or formula supplemented with long-chain polyunsaturates. Lipids 30, 941–948.

Makrides, M., Neumann, M.A., Simmer, K. and Gibson, R.A. (2000) A critical appraisal of the role of dietary long-chain polyunsaturated fatty acids on neural indices of term infants: a randomized, controlled trial. Pediatrics 105, 32–38.

Martin, R.E. (1998) Docosahexaenoic acid decreases phospholipase A2 activity in the neurites/nerve growth cones of PC12 cells. J. Neurosci. Res. 54, 805–813.

Martinez, M. (1990) Severe deficiency of docosahexaenoic acid in peroxisomal disorders: a defect of delta 4 desaturation? Neurology 40, 1292–1298.

Martinez, M. (1991) Developmental profiles of polyunsaturated fatty acids in the brain of normal infants and patients with peroxisomal diseases: severe deficiency of docosahexaenoic acid in Zellweger's and pseudo-Zellweger's syndromes. World Rev. Nutr. Diet. 66, 87–102.

Martinez, M. (1992) Treatment with docosahexaenoic acid favorably modifies the fatty acid composition of erythrocytes in peroxisomal patients. Prog. Clin. Biol. Res. 375, 389–397.

Martinez, M. (1996) Docosahexaenoic acid therapy in docosahexaenoic acid-deficient patients with disorders of peroxisomal biogenesis. Lipids 31, S145–S152.

Martinez, M. and Vazquez, E. (1998) MRI evidence that docosahexaenoic acid ethyl ester improves myelination in generalized peroxisomal disorders. Neurology 51, 26–32.

Martinez, M., Pineda, M., Vidal, R., Conill, J. and Martin, B. (1993) Docosahexaenoic acid – a new therapeutic approach to peroxisomal-disorder patients: experience with two cases. Neurology 43, 1389–1397.

McCreadie, R.G. (1997) The Nithsdale Schizophrenia Surveys 16. Breast-feeding and schizophrenia: preliminary results and hypotheses. Br. J. Psychiatry 176, 334–337.

McLennan, P., Howe, P., Abeywardena, M., Muggli, R., Raederstorff, D., Mano, M., Rayner, T. and Head, R. (1996) The cardiovascular protective role of docosahexaenoic acid. Eur. J. Pharmacol. 300, 83–89.

Minami, M., Kimura, S., Endo, T., Hamaue, N., Hirafuji, M., Togashi, H., Matsumoto, M., Yoshioka, M., Saito, H., Watanabe, S., Kobayashi, T. and Okuyama, H. (1997) Dietary docosahexaenoic acid increases cerebral acetylcholine levels and improves passive avoidance performance in stroke-prone spontaneously hypertensive rats. Pharmacol. Biochem. Behav. 58, 1123–1129.

Neuringer, M., Connor, W.E., Lin, D.S., Barstad, L. and Luck, S. (1986) Biochemical and functional effects of prenatal and postnatal omega 3 fatty acid deficiency on retina and brain in rhesus monkeys. Proc. Natl. Acad. Sci. U.S.A. 83, 4021–4025.

Neuringer, M., Anderson, G.J. and Connor, W.E. (1988) The essentiality of n-3 fatty acids for the development and function of the retina and brain. Annu. Rev. Nutr. 8, 517–541.

Phylactos, A.C., Harbige, L.S. and Crawford, M.A. (1994) Essential fatty acids alter the activity of manganese-superoxide dismutase in rat heart. Lipids 29, 111–115.

Phylactos, A.C., Leaf, A.A., Costeloe, K. and Crawford, M.A. (1995) Erythrocyte cupric/zinc superoxide dismutase exhibits reduced activity in preterm and low-birthweight infants at birth. Acta Paediatr. 84, 1421–1425.

Prasad, M.R., Lovell, M.A., Yatin, M., Dhillon, H. and Markesbery, W.R. (1998) Regional membrane phospholipid alterations in Alzheimer's disease. Neurochem. Res. 23, 81–88.

Rodriguez de Turco, E.B., Deretic, D., Bazan, N.G. and Papermaster, D.S. (1997) Post-Golgi vesicles cotransport docosahexaenoyl-phospholipids and rhodopsin during frog photoreceptor membrane biogenesis. J. Biol. Chem. 272, 10491–10497.

Salem Jr, N. and Niebylski, C.D. (1995) The nervous system has an absolute molecular species requirement for proper function. Mol. Membr. Biol. 12, 131–134.

Salem Jr, N. and Pawlosky, R.J. (1994) Arachidonate and docosahexaenoate biosynthesis in various species and compartments in vivo. World Rev. Nutr. Diet. 75, 114–119.

Salem Jr, N., Wegher, B., Mena, P. and Uauy, R. (1996) Arachidonic and docosahexaenoic acids are biosynthesized from their 18-carbon precursors in human infants. Proc. Natl. Acad. Sci. U.S.A. 93, 49–54.

Salem, N.J., Kim, H.-Y. and Yergey, J.A. (1986) Docosahexaenoic acid: membrane function and metabolism. In: Health Effects of Polyunsaturated Fatty Acids in Seafoods, Academic Press, Orlando, FL, pp. 263–317.

Sesler, A. and Ntambi, J. (1998) Polyunsaturated fatty acid regulation of gene expression. J. Nutr. 128, 923–926.

Simopoulos, A.P., Leaf, A. and Salem, N. (1999) Workshop on the essentiality of and recommended dietary intakes for omega-6 and omega-3 fatty acids. J. Am. Coll. Nutr. 18, 487–489.

Soderberg, M., Edlund, C., Kristensson, K. and Dallner, G. (1991) Fatty acid composition of brain phospholipids in aging and in Alzheimer's disease. Lipids 26, 421–425.

Stevens, L.J., Zentall, S.S., Deck, J.L., Abate, M.L., Watkins, B.A., Lipp, S.R. and Burgess, J.R. (1995) Essential fatty acid metabolism in boys with attention-deficit hyperactivity disorder. Am. J. Clin. Nutr. 62, 761–768.

Stoll, A.L., Severus, W.E., Freeman, M.P., Rueter, S., Zboyan, H.A., Diamond, E., Cress, K.K. and Marangell, L.B. (1999) Omega 3 fatty acids in bipolar disorder: a preliminary double-blind, placebo-controlled trial. Arch. Gen. Psychiatry 56, 407–412.

Stordy, B.J. (1995) Benefit of docosahexaenoic acid supplements to dark adaptation in dyslexics. Lancet 346, 385.

Stordy, B.J. (2000) Dark adaptation, motor skills, docosahexaenoic acid, and dyslexia. Am. J. Clin. Nutr. 71, 323S–326S.

Suh, M., Wierzbicki, A.A., Lien, E. and Clandinin, M.T. (1996) Relationship between dietary supply of long-chain fatty acids and membrane composition of long- and very long chain essential fatty acids in developing rat photoreceptors. Lipids 31, 61–64.

Suzuki, H., Manabe, S., Wada, O. and Crawford, M.A. (1997) Rapid incorporation of docosahexaenoic acid from dietary sources into brain microsomal, synaptosomal and mitochondrial membranes in adult mice. Int. J. Vitam. Nutr. Res. 67, 272–278.

Uauy, R., Birch, E., Birch, D. and Peirano, P. (1992) Visual and brain function measurements in studies of n-3 fatty acid requirements of infants. J. Pediatr. 120, S168–S180.

Vaddadi, K.S., Courtney, P., Gilleard, C.J., Manku, M.S. and Horrobin, F. (1989) A double-blind trial of essential fatty acid supplementation in patients with tardive dyskinesia. Psychiatr. Res. 27, 313–324.

Vanderhoof, J., Gross, S., Hegyl, T., Clandinin, T., Porcelli, P., De Christolano, J., Rhodes, T., Tsang, R., Shattuk, K., Cowett, R., Adamkin, D., McCarton, C., Heird, W., Hook-Morris, B., Pereira, G., Chan, G., Van Aerde, J., Boyle, F., Pramuk, K., Euler, A. and Lien, E. (1999) Evaluation of a long-chain polyunsaturated fatty acid supplemented formula on growth, tolerance, and plasma lipids in preterm infants up to 48 weeks postconceptual age. J. Pediatr. Gastroenterol. 29, 318–326.

Voss, A., Reinhart, M., Sankarappa, S. and Sprecher, H. (1991) The metabolism of 7,10,13,16,19-docosa-pentaenoic acid to 4,7,10,13,16,19-docosahexaenoic acid in rat liver is independent of a 4-desaturase. J. Biol. Chem. 266, 19995–20000.

Wei, J.W., Yang, L.M., Sun, S.H. and Chiang, C.L. (1987) Phospholipids and fatty acid profile of brain synaptosomal membrane from normotensive and hypertensive rats. Int. J. Biochem. 19, 1225–1228.

Willatts, P., Forsyth, J., DiModugno, M., Varma, S. and Colvin, M. (1998) Effect of long-chain polyunsaturated fatty acids in infant formula on problem solving at 10 months of age. Lancet 352, 688–691.

Xiao, Y.F., Gomez, A.M., Morgan, J.P., Lederer, W.J. and Leaf, A. (1997) Suppression of voltage-gated L-type Ca2+ currents by polyunsaturated fatty acids in adult and neonatal rat ventricular myocytes. Proc. Natl. Acad. Sci. U.S.A. 94, 4182–4187.

Xu, L.Z., Sanchez, R., Sali, A. and Heintz, N. (1996) Ligand specificity of brain lipid-binding protein. J. Biol. Chem. 271, 24711–24719.

Young, C., Gean, P.W., Wu, S.P., Lin, C.H. and Shen, Y.Z. (1998) Cancellation of low-frequency stimulation-induced long-term depression by docosahexaenoic acid in the rat hippocampus. Neurosci. Lett. 247, 198–200.

E.R. Skinner (Ed.), *Brain Lipids and Disorders in Biological Psychiatry*

CHAPTER 2

The effects of n-3 fatty acid deficiency and its reversal upon the biochemistry of the primate brain and retina

William E. Connor and Gregory J. Anderson

The Division of Endocrinology, Diabetes and Clinical Nutrition, Department of Medicine, L465, Oregon Health Sciences University, Portland, Oregon 97201

1. Introduction

The importance of dietary fatty acids in the structure and functioning of the brain and retina cannot be overemphasized (Connor et al. 1990). Membrane lipids constitute 50–60% of the solid matter in the brain (O'Brien 1986), and phospholipids are quantitatively the most significant component of membrane lipids (Crawford and Sinclair 1972). A major proportion of brain phospholipids contain long chain polyunsaturated fatty acids of the essential fatty acid classes, n-3 and n-6 (Crawford et al. 1976, O'Brien et al. 1964). These fatty acids usually occupy the sn-2 position of brain phospholipid molecules. Docosahexaenoic acid (22:6 n-3, DHA) is the predominant polyunsaturated fatty acid in the phospholipids of the cerebral cortex and retina. The primate brain gradually accumulates its full complement of DHA during intrauterine life and during the first year or more after birth (Clandinin et al. 1980a,b). DHA or its precursor n-3 fatty acids must be provided in the diet of the mother and infant for normal brain and retinal development.

In our previous reports, we have shown that infant rhesus monkeys born from mothers fed an n-3 fatty acid-deficient diet and then also fed a deficient diet after birth developed low levels of n-3 fatty acids in the brain and retina and impairment in visual function (Neuringer et al. 1984, 1986, Connor et al. 1984). The specific biochemical markers of the n-3 deficient state were a marked decrease in the DHA (22:6 n-3) of the cerebral cortex and a compensatory increase in n-6 fatty acids, especially docosapentaenoic acid (22:5 n-6). Thus, the sum total of the n-3 and n-6 fatty acids remained similar, about 50% of the fatty acids in phosphatidylethanolamine and phosphatidylserine, indicating the existence of mechanisms in the brain to conserve the polyunsaturation of membrane phospholipids as much as possible despite the n-3 fatty acid deficient state.

In this chapter, we report the results of two experiments in which we fed an n-3 fatty acid-rich repletion diet to n-3 deficient monkeys at two different time points: At birth (monkeys with prenatal deficiency) and at 12 to 24 months of age (prenatal and postnatal deficiency). In the first experiment in the newborn monkeys, the fatty acids of the brain were only partially assembled and repletion was attempted by feeding 18:3 n-3 (linolenic acid). In the second study in juvenile monkeys, as detailed below, the brain had time to reach biochemical maturity. Repletion of the n-3 fatty acid deficiency in the immature brain might be expected to occur more readily than in the mature brain.

In the second study, juvenile rhesus monkeys who had been fed an n-3 fatty acid deficient diet since intrauterine life were repleted with a fish oil diet rich in the n-3 fatty acids, DHA and 20:5 n-3 (eicosapentaenoic acid, EPA). In both studies the fatty acid composition was determined for the lipid classes of plasma and erythrocytes and for the phospholipid classes of frontal cortex samples obtained from serial biopsies and the time of autopsy. From these analyses, the half-lives of DHA and EPA in the phospholipids of plasma, erythrocytes, and cerebral cortex were estimated. In both groups of monkeys, the n-3 fatty acid deficient brain rapidly regained a normal or even supernormal content of DHA with a reciprocal decline in n-6 fatty acids, demonstrating that the fatty acids of the cortex turned over with relative rapidity under the circumstances of these experiments. Noteworthy is the fact that despite the different maturities of the brain at these two time points in development (birth and the juvenile state) and the use of two different diets, soy oil (containing 18:3 n-3) and fish oil (containing DHA and EPA), repletion of the n-3 fatty acid state seemed to occur similarly.

2. Study 1: Reversal of prenatal n-3 fatty acid deficiency

Five adult female rhesus monkeys were fed a semi-purified diet low in n-3 fatty acids for at least 2 months before conception and throughout pregnancy. The resulting infants were then fed a repletion diet, adequate in n-3 fatty acids, from birth until three years of age. The detailed composition of these semi-purified diets (Portman et al. 1967) is shown in Table 1. Safflower oil was used as the sole fat source for the deficient diet because it has a very low content of linolenic acid (18:3 n-3), 0.3% of total fatty acids and a very high ratio of n-6 to n-3 fatty acids (255 to 1). Soy oil comprised the repletion diet fed after birth as a liquid semi-purified formula. It had an excellent ratio of n-6 to n-3 fatty acids (7 to 1) and supplied linolenic acid (18:3 n-3) to the deficient infant monkeys at 7.7% of total fatty acids.

The fatty acids of the plasma, erythrocytes and cerebral cortex were then determined sequentially from birth onwards. In order to evaluate the degree of brain repletion, biopsies of the frontal cerebral cortex were obtained at 15–60 weeks of age from the prenatally n-3 fatty acid deficient monkeys as previously described (Connor et al. 1990). Biopsies constituted 15–30 mg samples of pre-frontal cortex gray matter obtained through a small burr hole in the frontal bone of the skull under thiamytal anesthesia (25 mg/kg). No behavioral or neurological changes were noted after the biopsies. A total of 3–4 biopsies were performed on each monkey. All monkeys were killed at 3 years of age. The plasma, erythrocytes, brain and retina were collected for analysis. The methods of biochemical analysis were as reported previously (Connor et al. 1990).

As this study was designed to determine the long-term effect of an exclusively "prenatal n-3 fatty acid deficiency", the infant monkeys were subsequently fed a soy oil control diet from birth to termination at 3 years of age. Feeding an n-3 fatty acid deficient to pregnant rhesus monkeys produced a severe n-3 fatty acid deficiency in infant monkeys at birth. The deficiency at birth was severe in all tissues examined, but was particularly pronounced in the plasma (Table 2). Plasma n-3 fatty acids, including DHA, were depleted by 86% relative to control. Although 22:5 n-6, the classic indicator of tissue n-3 fatty

Table 1
The fatty acid composition of the experimental diets (% of total fatty acids)

Fatty acid	Prenatal n-3 fatty acid deficient diet (safflower oil)	Prenatal control diet (n-3 fatty acid adequate) (soy oil)	Postnatal control and repletion diet (n-3 fatty acid adequate)[a] (soy oil)	Fish oil repletion diet to n-3 fatty acid deficient animals
16:0	7.1	10.7	10.7	15.8
18:0	2.5	4.2	4.2	3.1
Saturated	10.0	16.4	16.4	27.7
18:1 (n-9)	13.3	23.7	23.7	13.0
Monounsaturated	13.8	24.2	24.2	28.4
18:2 (n-6)	76.0	53.1	53.1	1.7
20:4 (n-6)	–	–	–	1.1
Total (n-6)	76.5	53.4	53.4	3.6
18:3 (n-3)	0.3	7.7	7.7	0.7
20:5 (n-3)	–	–	–	16.4
22:6 (n-3)	–	–	–	11.0
Total (n-3)	0.3	7.7	7.7	34.2
n-6/n-3	255.0	7.0	7.0	0.1

[a] Fed at birth to both the control infants and the n-3 fatty acid prenatal deficient infants.

acid deficiency, was only mildly increased in the plasma at birth, the depleted n-3 fatty acids in plasma were mostly substituted by palmitic acid (16:0). After 3 years consuming the soy oil diet, the formerly n-3 fatty acid deficient monkeys showed no signs of the deficiency in the plasma fatty acid composition. In fact, plasma n-3 fatty acids returned to normal within 4 weeks of birth [6.0 ± 2.4% vs. 6.1 ± 1.2 (control). To be commented on subsequently was the marked decline in plasma DHA after birth in the control monkeys fed the soy oil diet, which supplied only the precursor fatty acid 18:3 n-3. Their mother had received the same soy diet during pregnancy. Erythrocyte fatty acids responded to the intrauterine n-3 fatty acid deficient diet in a fashion similar to plasma fatty acids. At birth, however, the depleted n-3 fatty acids were substituted mostly by the long-chain n-6 fatty acids, 22:4 n-6 and 22:5 n-6. After eight weeks of the soy oil diet, the n-6 fatty acids had declined greatly and the n-3 fatty acids in erythrocytes had recovered to control values (Fig. 1). At 3 years of age only a trace of the deficiency remained. Note also in Fig. 1 the marked decline in the DHA of the control monkeys, from 13.7% of total fatty acids at birth to 6.3% at 10 weeks of age, the same level of DHA found in the repleted monkeys also fed soy oil. This decline in blood DHA levels is also seen in human infants fed a standard formula containing soy oil as a source of the n-3 fatty acid linolenic acid but not seen in infants fed human milk or formulas enriched with DHA (Austed et al. 1997).

The phospholipids of newborn infant brain and retina, normally highly enriched in

Table 2
The fatty acid composition of plasma phospholipids from prenatally n-3 deficient monkeys at birth and after 3 years of repletion with a soy oil diet (% of total fatty acids, mean ± SD)

Fatty acid	Deficient diet (safflower oil) at birth ($n=4$)	Control diet (soy oil) at birth ($n=6$)	Repletion diet (soy oil) after 3 yr ($n=3$)	Control diet (soy oil) after 3 yr ($n=3$)
16:0	30.2±2.7	16.3±11.3	18.0±1.3	16.7±0.2
18:0	15.6±0.4	16.4±2.6	16.6±2.4[a]	21.2±0.9[b]
Saturated	47.1±3.0[a]	33.3±9.9	35.8±4.0[b]	39.3±0.9
18:1	9.0±0.9	10.2±2.7	7.8±0.3	7.6±0.4
Monounsaturated	10.8±1.9	11.3±3.1	8.9±1.3	9.0±0.9
18:2(n-6)	21.6±4.4	23.1±6.4	37.6±7.2[b]	38.6±2.3[b]
20:4(n-6)	13.6±4.7	14.4±4.8	6.6±4.4	5.8±1.5[b]
22:4(n-6)	1.0±0.5	0.5±0.4	0.7±0.3	0.4±0.1
22:5(n-6)	1.8±0.6[a]	0.4±0.3	0.8±0.3[a,b]	0.1±0.1
Total (n-6)	40.3±4.5	43.9±6.6	49.4±0.4[b]	47.6±0.3
18:3(n-3)	0.1±0.2	0.3±0.4	1.1±0.7[b]	0.4±0.1
22:5(n-3)	0.1±0.1	0.6±0.3	0.8±0.4	1.0±0.2
22:6(n-3)	1.5±0.5[a]	10.2±6.4	1.7±1.3	1.5±0.9[b]
Total(n-3)	1.6±0.7[a]	11.1±6.3	4.0±2.8	3.4±1.4[b]

[a] Different from control at respective age, $P < 0.03$.
[b] Different from birth for respective diet, $P < 0.03$.

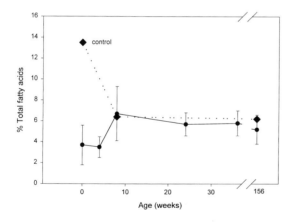

Fig. 1. Repletion of n-3 fatty acids in erythrocytes of prenatally n-3 fatty acid deficient monkeys fed a control diet for 3 years after birth. Solid diamonds: control values.

DHA, were severely depleted of DHA and other long-chain n-3 fatty acids after the pregnant monkeys had been fed the deficient diet. For the most part, the n-6 fatty acid 22:5 n-6 substituted for DHA. After three years on the repletion diet, DHA in the phosphatidylethanolamine (PE) of the frontal cortex increased from 3.5% of total fatty acids to 26.6% (control at 3 years was 23.0%). These monkeys showed only a trace of n-3 fatty acid deficiency with the possible exception of continued low DHA

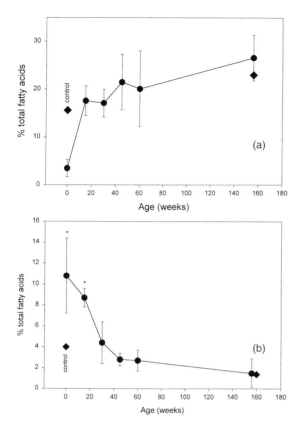

Fig. 2. (a) Repletion of docosahexaenoic acid (DHA) in brain PE of prenatally n-3 fatty acid deficient monkeys fed a control diet for 3 years after birth. Data at 15–60 weeks are from brain biopsies. Solid diamonds: control values. (b) Disappearance of 22:5 n-6 in brain PE of prenatally n-3 fatty acid deficient monkeys fed a control diet for 3 years after birth. Data at 15–60 weeks are from brain biopsies. Solid diamonds: control values. $^*P = 0.02$ vs. 0-week control.

in phosphatidylethanolamine of the retina, but still displayed elevated levels of 22:5 n-6. In this case, recovery was substantial but may have been incomplete. The data for DHA of phosphatidylserine in brain was similar with an increase from 5.1% of total fatty acids to 30.5% after the repletion diet.

Recovery from n-3 deficiency in the brain was relatively rapid. Serial brain biopsies of recovering infants revealed that 22:6 n-3 levels in brain phosphatidylethanolamine had returned to control values by the time of the earliest point sampled, namely 15 weeks of age (Fig. 2a). On the other hand, levels of 22:5 n-6 in brain phosphatidylethanolamine did not decline to control values until 30 weeks (Fig. 2b). At 15 weeks of age the level of 22:5 n-6 in brain PE was still more than two times greater than control values and it took about 25 weeks for this n-6 fatty acid to decline to control values (Fig. 2b). The lag in the fall of 22:5 n-6 is noteworthy.

3. Study 2: Repletion of n-3 fatty acids in deficient juvenile monkeys

Juvenile rhesus monkeys who had developed n-3 fatty acid deficiency since intrauterine life and had been maintained on the same deficient diet for 10–24 months after birth

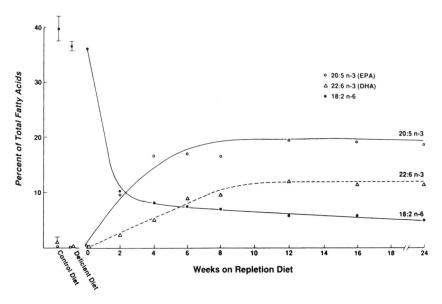

Fig. 3. The time course of mean fatty acid changes in plasma phospholipids after the feeding of fish oil. Note that reciprocal changes of the two major n-3 (EPA and DHA) and the major n-6 (18:2) polyunsaturated fatty acids occurred as n-3 fatty acids increased and n-6 fatty acids decreased. The concentrations of these fatty acids in the plasma phospholipids of monkeys fed the control soybean oil and safflower oil diet from our previous study (Neuringer et al. 1986) are given for comparison. Expressed as percentage of total fatty acids, DHA in control monkeys was 1.1±0.7%; EPA, 0.2±0.1%; 18:2 n-6, 39.6±2.3%. In deficient monkeys, DHA was 0%, 18:2 n-6, 36.7±0.7%.

were then repleted with a fish oil diet rich in the n-3 fatty acids, DHA and 20:5 n-3 (eicosapentaenoic acid, EPA). The fatty acid compositions were determined for the lipid classes of plasma and erythrocytes and from the phospholipid classes of frontal cortex samples obtained from serial biopsies and at the time of autopsy. From these analyses, the half-lives of DHA and EPA in the phospholipids of plasma, erythrocytes, and cerebral cortex were estimated. The deficient animal rapidly regained a normal or even supernormal content of DHA with a reciprocal decline in n-6 fatty acids, demonstrating that the fatty acids of the blood and the gray matter of the brain turned over with relative rapidity under the circumstances of these experiments.

The establishment of the n-3 fatty acid deficiency was documented by the biochemical changes, electroretinographic abnormalities, and visual acuity loss, as described previously in these monkeys before repletion (Neuringer et al. 1984, 1986, Connor et al. 1984). Beginning at 10 to 24 months of age, the five juvenile monkeys were then given the same semi-purified diet with fish oil replacing 80% of the safflower oil as the fat source. The remaining 20% safflower oil provided ample amounts of n-6 fatty acids as linoleic (18:2 n-6) at 4.5% of calories.

Changing from the n-3 fatty acid-deficient diet (safflower oil) to the n-3-rich diet (fish oil) increased the total plasma n-3 fatty acids greatly, from 0.1 to 33.6% of total fatty acids (Fig. 3). EPA, which was especially high in the fish oil, contributed the major increase, from zero to 22.1%, and represented 66% of the total n-3 fatty acid increase.

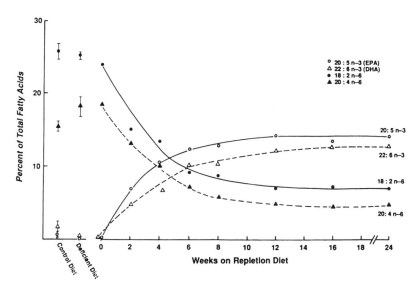

Fig. 4. The time course of mean fatty acid changes in erythrocyte phospholipids after the feeding of fish oil. Note that reciprocal changes of the two major n-3 (EPA and DHA) and the two major n-6 (18:2 and 20:4) polyunsaturated fatty acids occurred as n-3 fatty acids increased and n-6 fatty acids decreased. The concentrations of these four fatty acids in the erythrocyte phospholipids of monkeys fed the control soybean oil and the deficient safflower oil diet from our previous study (Neuringer et al. 1986) are given for comparison. Expressed as percentage of total fatty acids, DHA in control monkeys was 1.7±0.9%; EPA, 0.5±0.1%; 20:4 n-6, 15.4±0.5%; 18:2 n-6, 25.7±1.0%. In deficient monkeys, DHA was 0.2±0%; EPA, 0%; 20:4 n-6, 18:2±1.3%; 18:2 n-6 25.2±0.4%.

DHA increased from 0.1 to 8.3% and 22:5 n-3 from zero to 2.1%. A major reciprocal decrease occurred in the n-6 fatty acid linoleic acid, which was reduced from 54.3 to 9.2% of total fatty acids while total n-6 fatty acids fell from 65.4 to 15.5%. The change in arachidonic acid, however, was relatively small, from 6.4 to 5.5%. These changes were certainly manifest as early as 2 weeks after repletion and were completed by 6–8 weeks. The restored fatty acid values then remained constant until autopsy.

The four major plasma lipid fractions (phospholipids, cholesteryl esters, triglycerides, and free fatty acids) exhibited similar changes in response to the fish oil diet. In the phospholipid fraction, the n-3 fatty acids increased from 0.4% to 34% in the fish oil diet. EPA increased from zero to 19%, accounting for 55% of the total increase. Linoleic acid reciprocally decreased from 36 to 5% and total n-6 fatty acids from 48 to 12% of total fatty acids. In cholesteryl esters, n-3 fatty acids increased from 0.2 to 36%. An increase in EPA from zero to 31% accounted for 86% of the increase, a much greater proportion than in phospholipids, whereas DHA only increased from zero to 4%. The decline in n-6 fatty acids from 77 to 23% was largely accounted for by a decrease in linoleic acid from 73 to 17%. Similar changes were seen in the triglycerides and free fatty acid fractions.

The major unsaturated fatty acids of the phospholipids of erythrocytes from before fish oil feeding to the time of the last cortical biopsy (after 12–28 weeks of fish oil feeding) are displayed in Fig. 4. The total n-3 fatty acids increased from 1.3% to 31% of total fatty acids after the fish oil diet. EPA and DHA had similar increases, from 0.2 to 14%

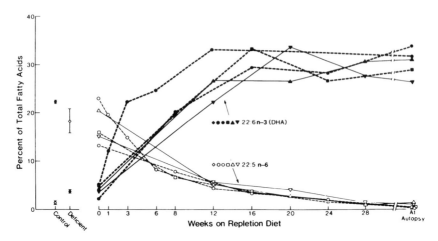

Fig. 5. The time course of fatty acid changes in phosphatidylethanolamine of the cerebral cortex of five juvenile monkeys fed fish oil for 43–129 weeks. As DHA increased, 22:5 n-6 decreased reciprocally. Levels of DHA and 22:5 n-6 in phosphatidylethanolamine of the frontal cortex of monkeys fed control (soybean oil) and deficient diets from a previous study (Neuringer et al. 1986) are given for comparison. DHA and 22:5 n-6 in control monkeys were 22.3 ± 0.3 and $1.4\pm0.3\%$ of total fatty acids, respectively. DHA and 22:5 n-6 in the deficient monkeys were 3.8 ± 0.4 and $18.3\pm2.5\%$, respectively.

for EPA and from 0.2 to 13% for DHA. In erythrocytes the n-6 fatty acids showed a significant decrease which occurred reciprocally. Arachidonic acid fell from 18 to 6% and linoleic acid declined from 24 to 7%. The n-3 fatty acids (EPA, DHA) after the fish oil feeding increased greatly from almost zero to about 14% of total fatty acids. The fatty acid composition of erythrocytes reached a new steady state by 12 weeks after fish oil feeding and remained constant until autopsy, but marked changes in fatty acid composition were already extensive after 2 weeks of the repletion diet.

Dramatic changes in the fatty acids of the frontal cortex were detected within 1 week after the fish oil diet was given as demonstrated in the individual data in the frontal lobe biopsy specimens from five juvenile monkeys. All four major phospholipid classes of the brain underwent extensive remodeling of their constituent fatty acids. The data in Fig. 5 is for the fatty acids of phosphatidylethanolamine from each of the five experimental monkeys. By 12–28 weeks, the total n-3 fatty acids increased from 4 to 36% of total fatty acids (Connor et al. 1990). The major increase was in DHA, from 4 to 29%, while EPA and 22:5 n-3, another n-3 fatty acid found in fish oil each increased from zero to almost 3%. To be emphasized, as will be discussed later, is the apparent conversion of EPA to DHA in the brain. The total n-6 fatty acids reciprocally decreased from 44 to 16% of the total fatty acids, with the major reduction occurring in 22:5 n-6, from 18 to 2%, and 22:4 n-6, from 12 to 4%. There was also a moderate decrease of arachidonic acid from 12.8 to 8.9% of total fatty acids. Again, a major remodeling of the phospholipid fatty acids from n-6 to n-3 fatty acids was evident.

In the phosphatidylserine (PS) fraction, the total n-3 fatty acids also increased greatly from 4.8 to 36% of total fatty acids with DHA increasing from 5 to 32%. Total n-6 fatty acids decreased from 38 to 11%. Again, the major and reciprocal decrease was in 22:5 n-6, from 20 to 4%. In the phosphatidylinositol (PI) fraction, total n-3 fatty acids increased form 3 to 18%, while n-6 fatty acids decreased from 43 to 25% and 22:5 n-6 decreased from 11 to 0.7%. Although the content of n-3 fatty acids in phosphatidylcholine (PC) is relatively small even in the normal brain, this fraction also showed an increase in the n-3 fatty acids from 0.4 to 5.1% after fish oil feeding. Unlike PE and PS, the fish oil feeding did not decrease the arachidonic acid levels in PI and PC.

In summary, fish oil feeding resulted in reciprocal changes in the levels of n-3 and n-6 fatty acids in the phospholipids of cerebral cortex. The two major 22-carbon n-3 and n-6 fatty acids, DHA and 22:5 n-6, were responsible for the greatest changes. Figure 5 plots the changes in these two fatty acids in phosphatidylethanolamine, plus the analogous values for juvenile monkeys fed a control (soybean oil) diet and deficient (safflower oil) diet from our previous study (Neuringer et al. 1986). In control monkeys the DHA and 22:5 n-6 contents of frontal cortex were 22.3 ± 0.3 and $1.4 \pm 0.3\%$ of total fatty acids respectively, whereas in deficient monkeys they were nearly the reverse, $3.8 \pm 0.4\%$ DHA and $18.3 \pm 2.5\%$ 22:5 n-6 (Neuringer et al. 1986). Reflecting the high content of the long chain n-3 fatty acids of fish oil, the DHA content of cerebral cortex in the fish oil monkeys became even higher than in the soybean oil-fed control monkeys (29.3 ± 2.6 vs. $22.3 \pm 0.3\%$, $p < 0.025$).

4. Turnover of fatty acids in various tissues

Using the serial data for the cerebral cortex, plasma, and erythrocytes, we constructed accumulation and decay curves for several key fatty acids in these tissues, which provided gross estimates of their turnover times after fish oil feeding to n-3 fatty acid-deficient monkeys (Table 3). For cerebral cortex, a steady state was reached after 12 weeks of fish oil feeding for DHA, but 22:5 n-6 took longer to decline to the low levels found in the cortex of control animals. The half-lives of DHA in cerebral phospholipids ranged from 17 to 21 days: 21 days for phosphatidylethanolamine, 21 days for phosphatidylserine, 18 days for phosphatidylinositol, and 17 days for phosphatidylcholine. The corresponding values for 22:5 n-6 in these same phospholipids were 32, 49, 14 and 28 days, respectively. The half-lives of linoleic acid, EPA and DHA in plasma phospholipids were estimated to be 8, 18, and 29 days, respectively. In the phospholipids of erythrocytes, linoleic acid, arachidonic acid, EPA and DHA had half-lives of 28, 32, 14 and 21 days, respectively.

5. Discussion

5.1. General

Before 1940 it was generally considered that phospholipids, once laid down in the nervous system of mammals during growth and development, were comparatively static entities. However, later studies using (^{32}P)orthophosphate showed that brain phospholipids as a whole are metabolically active in vivo (Dawson and Richter 1950, Ansell and Dohmen 1957). In the present study, by following the changes in phospholipid fatty acid

Table 3
The turnover of fatty acids in various tissues

Fatty acid	Half-life (days)		
	DHA	EPA	22:5 n-6
Plasma phospholipids	29	18	
Erythrocyte phospholipids	21	14	
Cerebral cortex			
Phosphatidylethanolamine	21		32
Phosphatidylserine	21		49
Phosphatidylinositol	18		14
Phosphatidylcholine	17		28

composition, we have demonstrated that an n-3 fatty acid-enriched diet can rapidly reverse a severe n-3 fatty acid deficiency in the brains of primates. The phospholipid fatty acids of the cerebral cortex of juvenile monkeys are in a dynamic state and are subject to continuous turnover under certain defined conditions.

The observed changes in fatty acid composition could be the result of a complete breakdown and re-synthesis of cortical phospholipids or a turnover of only the fatty acids in the *sn*-2 position, which has the higher proportion of polyunsaturated acyl groups. Turnover of fatty acids in the *sn*-2 position is well known, and is commonly referred to as deacylation/reacylation (Corbin and Sun 1978, Fisher and Rowe 1980). This process could have an important role in maintaining optimal membrane composition without the high-energy cost associated with de novo phospholipid synthesis.

The reversibility of n-3 fatty acid deficiency in the monkey cerebral cortex was relatively rapid in our study. Effects of fish oil feeding were seen within 1 week after its initiation. By that time, DHA in the phosphatidylethanolamine of the cerebral cortex had more than doubled. The DHA concentration in phosphatidylethanolamine reached the control value of 22% in 6–12 weeks after fish oil feeding. We and others have demonstrated that the uptake of DHA and other fatty acids occurs within minutes after their intravenous injection bound to albumin (Anderson and Connor 1988). Furthermore, DHA is taken up by the brain in preference to other fatty acids (Anderson and Connor 1988). In contrast to the rapid incorporation of DHA, 22:5 n-6 only decreased from 23% to 20% during the first week. This asymmetry may indicate that DHA did not exclusively displace 22:5 n-6 from the *sn*-2 position of brain phospholipids.

In the present study, the half-life of DHA of phosphatidylethanolamine in cerebral cortex was similar to the half-lives of DHA in plasma and erythrocyte phospholipids, roughly 21 days. These data suggest that the blood–brain barrier present for cholesterol (Pitikin et al. 1972, Olendorf 1975) and other substances may not exist for the fatty acids of the plasma phospholipids because of the relatively rapid uptake of plasma DHA into the brain. The mechanisms of transport of these fatty acids remain to be investigated.

Similar reversals of biochemical deficiencies of n-3 fatty acids or of total essential fatty acids have been studied in the rodent brain under somewhat different experimental conditions (Trapp and Bernson 1977, Walker 1967, Sanders et al. 1984, Youyou et al. 1986, Homayoun et al. 1988, Odutuga 1981). In a study by Youyou et al. (1986), complete recovery from the n-3 fatty acid deficiency, as measured by an increase of DHA and a decrease of 22:5 n-6, required 13 weeks as compared to 6–12 weeks in our monkeys. There were major differences between this study and ours, which may be responsible for the different recovery rates observed. Most importantly we fed monkeys fish oil, which is high in DHA and EPA, whereas they fed young rats soy oil, which is lower in total n-3 fatty acids, and contains only the precursor of DHA, linolenic acid (18:3 n-3). The in situ biosynthesis of DHA from 18:3 n-3 may be a rate-limiting factor because of low desaturase activity, and also because the precursor 18:3 n-3 is oxidized much more readily than DHA (Leyton et al. 1987). Dietary 18:3 n-3 is also less effective in promoting biochemical recovery in n-3 fatty acid deficient chicks than dietary DHA or EPA (Anderson et al. 1989). Indeed, in our repletion studies reported here, dietary 18:3 n-3 was also much slower than dietary EPA/DHA in achieving repletion. By four weeks after beginning repletion with EPA/DHA, the erythrocyte and brain already showed substantial recovery. On the other hand, monkeys repleted with 18:3 n-3 showed no change in erythrocyte n-3 fatty acids at this time. Since erythrocyte and brain DHA levels in repleting monkeys are closely connected (Connor et al. 2002), we can infer that brain n-3 fatty acids had also not yet begun to recover at this time-point. Ultimately, however, dietary 18:3 n-3 was effective at reversing the n-3 fatty acid deficiency. The effects of dietary n-3 fatty acids from fish oil, linseed oil, and soy oil upon the lipid composition of the rat brain have also been reported by several groups of investigators (Cocchi et al. 1984, Tarozzi et al. 1984, Carlson et al. 1986, Philbrick et al. 1987, Hargreaves and Clandinin 1988).

Because of the reciprocal changes of n-3 and n-6 fatty acids with fish oil feeding, the sum total of n-3 and n-6 polyunsaturation of the brain of animals fed n-3 deficient and fish oil diets remained very similar. However, the unsaturation index of phospholipids of the frontal cortex was higher in the fish oil-fed monkeys than in n-3 deficient monkeys. The functional significance of this difference in the unsaturation index is not known. Since phospholipids rich in polyunsaturated fatty acids constitute an integral part of brain and retinal membranes, the degree of unsaturation of these fatty acids may have an important influence on the structure of the membranes and their functions, via changes in biophysical properties and/or the activities of membrane-bound proteins including enzymes, receptors, or transport systems (Op den Kamp et al. 1985).

In our n-3 deficient monkeys, the electroretinograms showed several abnormalities before fish oil was fed (Neuringer et al. 1986). After repletion, when the concentration of DHA had been restored to above normal levels in the brain and retina, the electroretinograms remained abnormal (Connor and Neuringer 1988). The reason for the failure of the electroretinogram to improve is unknown but may relate to the time of repletion in the animal's development or to the use of fish oil containing EPA as well as DHA. The use of purified DHA, when it is available in quantity, might be a more physiologic way of repleting n-3 fatty acid deficient monkeys. As demonstrated in the present experiment, fish oil feeding was able not only to reverse the n-3 fatty acid-deficient state in the brain, but also to increase the n-3 acid content in brain phospholipids above control levels. Furthermore, EPA increased from zero in control monkeys to 3.1% after fish oil feeding and 22:5 n-3 increased from 0.1 to 3.5%. At the same time, arachidonic acid and 22:4 n-6 decreased to below control levels. Whether this "overload" of n-3 fatty acids and perhaps un-physiologic reduction of n-6 fatty acids in brain phospholipids was advantageous or detrimental in terms of membrane function is uncertain. Similar considerations would apply to the retina and its functioning.

An interesting question relates to the blood and brain ratios of EPA and DHA. The concentrations of EPA were high in the plasma and low in the brain, with DHA high in both brain and blood. For the erythrocytes the EPA /DHA was 1.0; for the brain the ratio was 0.12. It is clear that EPA was metabolized, in contrast to DHA, between the blood compartment and the brain. EPA could have entered the brain and there been converted to DHA by astrocytes which do have that capacity (Moore et al. 1991). Less likely it could have been catabolized with one pathway leading to the synthesis of the series 3 prostaglandins (Yerram et al. 1989).

5.2. Applicability to humans

What are the implications of these studies of n-3 fatty acid deficiency and subsequent repletion in rhesus monkeys to human beings? First of all, these data strongly suggest that any n-3 fatty acid deficient state will be corrected by a diet containing n-3 fatty acids from soy oil (18:3 n-3 linolenic acid) or fish oil containing EPA and DHA. The brain phospholipids will readily assemble the correct amounts of DHA in the *sn*-2 position of the phospholipid molecular species. Furthermore, other fatty acids of the n-6 series, which occupy that position in the deficient state, will ultimately be removed and replaced by DHA. It is not certain yet if functional abnormalities would likewise be corrected since the appropriate function may need to take place at a certain stage of development or it may not occur at all or in a lesser degree.

A second point is that the diagnosis of an n-3 fatty acid deficiency state in the tissues, especially the brain and retina, can well be approximated by analysis of plasma and red blood cells. This is particularly true if the deficient state occurs during pregnancy and from birth onwards.

A third point is that the brain presumably has the capacity to synthesize DHA from precursor forms such as EPA. The presumption of this statement is based on the fact that high plasma and red blood cell concentrations of EPA are not mirrored by similar concentrations in the brain; instead it is DHA that predominates so greatly. The alternative explanation is that EPA is rapidly metabolized once it enters the brain, but there is no evidence to support this particular point. Lastly, normal monkeys fed 18:3 n-3 from soy oil have a pronounced fall in DHA concentrations from birth onwards and deficient monkeys repleted with soy oil have increased DHA concentration but not more than the normal soy oil fed monkeys. This same situation prevails in humans in that soy oil formula-fed human infants likewise have a fall in blood DHA concentrations that does not occur in breast-fed infants or infants fed formulas supplemented with DHA. These data suggest that both the shorter chain n-3 fatty acids, e.g., 18:3 n-3, and the longer-chain n-3 fatty acids, e.g., DHA and EPA, will correct the n-3 fatty acid deficiency. In all probability, however, the EPA and DHA from fish oil apply a more suitable correction and are perhaps more physiological. Pure DHA might be even better.

Several questions are raised by the rapid incorporation of dietary DHA and other n-3 fatty acids from fish oil into phospholipid membranes of the cerebral cortex of juvenile rhesus monkeys. Would primate brains of "normal" fatty acid composition incorporate dietary DHA just as avidly as the brains of n-3-deficient monkeys? This situation is analogous to humans consuming large quantities of fish oil, and has been tested in adult rats (Bourre et al. 1988), where a 30% rise in brain DHA was observed. Would such a change in the primate brain be deleterious or advantageous? Quantities of EPA in the erythrocytes and cerebral cortex of the fish oil-supplemented monkeys were much higher than is normally the case. These abnormal levels might lead to functional disturbances, but no information is available about this point. Future studies of fish oil feeding to normal adult monkeys may provide answers to these questions, especially if molecular species of fatty acids of the phospholipid classes are determined and if the function of the "changed" organs is measured. For example, when we analyzed individual phospholipid molecular species of the brains of monkeys fed different diets, we observed highly significant dietary effects (Lin et al. 1990).

If DHA turns over as rapidly in the adult normal brain as in the deficient monkey brain, then perhaps the brain should be provided with a constant supply of DHA or other n-3 fatty acids. Ultimately, dietary sources of n-3 fatty acids would be desirable in both adults and infants (Neuringer et al. 1988). Whether the n-3 fatty acid supplied from the diet should be as 18:3 n-3 or preformed DHA or both is not completely known. It is possible that ample amounts of DHA could be synthesized from 18:3 n-3 via the desaturation and elongation pathways. However, the active uptake of DHA by the infant rat brain over other fatty acids suggests preference for acquiring preformed DHA directly from the blood (Olendorf 1975). In view of the significant impact of diet on brain composition, it will be important in the future to address the question of the appropriate amount and type of dietary n-3 fatty acids for optimal brain development during infancy and for maintenance during adult life (Neuringer and Connor 1986).

Acknowledgments

This work was supported by the Oregon Health Sciences Foundation. We are grateful to Michelle Burke for the manuscript preparation.

References

Anderson, G.J. and Connor, W.E. (1988) Uptake of fatty acids by developing rat brain. Lipids 23, 286–290.

Anderson, G.J., Connor, W.E. and Corliss, J.D. (1989) Docosahexaenoic acid is the preferred dietary n-3 fatty acid for the development of the brain and retina. Pediatr. Res. 27, 89–97.

Ansell, G.B. and Dohmen, H. (1957) The metabolism of individual phospholipids in rat brain during hypoglycemia, anesthesia and convulsions. J. Neurochem. 2, 1–10.

Austed, N., Montalto, M.B., Hall, R.T., Fitzgerald, K.M., Wheeler, R.E., Connor, W.E., Neuringer, M., Connor, S.L., Taylor, J.A. and Hartman, E.E. (1997) Visual acuity, erythrocyte fatty acid composition and growth in term infants fed formulas with long chain polyunsaturated (LCP) fatty acids for one year. Pediatr. Res. 41, 1–10.

Bourre, J.M., Bonneil, M., Dumont, O., Piciotti, M., Nalbone, G. and Lafont, H. (1988) High dietary fish oil alters brain polyunsaturated fatty acid composition. Biochim. Biophys. Acta 960, 458–461.

Carlson, S.E., Carver, J.D. and House, S.G. (1986) High fat diets varying in ratios of polyunsaturated fatty acid and linoleic to linolenic acid: a comparison of rat neural and red cell membrane phospholipids. J. Nutr. 11, 718–725.

Clandinin, M.T., Chappell, J.E., Leong, S., Heim, T., Syter, P.R. and Chance, G.W. (1980a) Intrauterine fatty acid accretion rates in human brain: implications for fatty acid requirements. Early Hum. Dev. 4, 121–129.

Clandinin, M.T., Chappell, J.E., Leong, S., Heim, T., Syter, P.R. and Chance, G.W. (1980b) Extrauterine fatty acid accretion rates in human brain: implications for fatty acid requirements. Early Human Dev. 4, 131–138.

Cocchi, M.C., Pignatti, M., Carpigiani, G., Tarozzi, G. and Turchetto, E. (1984) Effect of C18:3 (n-3) dietary supplementation on the fatty acid composition of the rat brain. Acta Vitaminol. Enzymol. 6, 151–156.

Connor, W.E. and Neuringer, M. (1988) The effects of n-3 fatty acid deficiency and repletion upon the fatty acid composition and function of the brain and retina. In: M.L. Karnowsky, A. Leaf and L.C. Bolis (Eds.), Biological Membranes: Aberrations in Membrane Structure and Function, Alan R. Liss, New York, pp. 275–294.

Connor, W.E., Neuringer, M., Barstad, L. and Lin, D.S. (1984) Dietary deprivation of linolenic acid in rhesus monkeys: effects on plasma and tissue fatty acid composition and on visual function. Trans. Assoc. Am. Physicians 97, 1–9.

Connor, W.E., Neuringer, M.A. and Lin, D.S. (1990) Dietary effects on brain fatty acid composition: the reversibility of n-3 fatty acid deficiency and turnover of docosahexaenoic acid in the brain, erythrocytes, and plasma of rhesus monkeys. J. Lipid Res. 31, 237–247.

Connor, W.E., Lin, D.S. and Neuringer, M. (2002) Will the docosahexaenoic acid content in the erythrocytes predict brain docosahexaenoic acid? Manuscript in preparation.

Corbin, D.R. and Sun, G.Y. (1978) Characterization of the enzymic transfer of arachidonoyl groups to 1 acyl-phosphoglycerise in mouse synaptosome fraction. J. Neurochem. 30, 77–82.

Crawford, M.A. and Sinclair, A.J. (1972) Nutritional influences in evolution of mammalian brain. In: Lipids, Malnutrition and the Developing Brain, Ciba Foundation Symposium, Associated Scientific Publishers, Amsterdam, pp. 267–287.

Crawford, M.A., Casperd, N.M. and Sinclair, A.J. (1976) The long chain metabolites of linoleic and linolenic acids in liver and brain in herbivores and carnivores. Comp. Biochem. Physiol. 54B, 395–401.

Dawson, R.M.C. and Richter, D. (1950) The phosphorous metabolism of the brain. Proc. R. Soc. B 137, 252–267.

Fisher, S.K. and Rowe, C.E. (1980) The acylation of lysophosphatidylcholine by subcellular fractions of guinea pig cerebral cortex. Biochim. Biophys. Acta 618, 231–241.

Hargreaves, K.M. and Clandinin, M.T. (1988) Dietary control of diacylphosphatidylethanolamine species in brain. Biochim. Biophys. Acta 962, 98–104.

Homayoun, P., Durand, G., Pascal, G. and Bourre, J.M. (1988) Alteration in fatty acid composition of adult rat brain capillaries and choroid plexus induced by a diet deficient in n-3 fatty acids: slow recovery after substitution with a non-deficient diet. J. Neurochem. 51, 45–48.

Leyton, J., Drury, P.J. and Crawford, M.A. (1987) Differential oxidation of saturated and unsaturated fatty acids in vivo in the rat. Br. J. Nutr. 57, 383–393.

Lin, D.S., Connor, W.E., Anderson, G.J. and Neuringer, M. (1990) The effect of diet upon the phospholipid molecular species composition of monkey brain. J. Neurochem. 55, 1200–1207.

Moore, S.A., Yoder, E., Murphy, S., Dutton, G.R. and Spector, A.A. (1991) Astrocytes, not neurons, produce docosahexaenoic acid (22:6 omega-3) and arachidonic acid (20:4 omega-6). J. Neurochem. 56, 518–524.

Neuringer, M. and Connor, W.E. (1986) Omega-3 fatty acids in brain and retina: evidence for their essentiality. Nutr. Rev. 44, 285–294.

Neuringer, M., Connor, W.E., Van Patten, C. and Barstad, L. (1984) Dietary omega-3 fatty acid deficiency and visual loss in infant rhesus monkeys. J. Clin. Invest. 73, 272–276.

Neuringer, M., Connor, W.E., Lin, D.S., Barstad, L. and Luck, S.J. (1986) Biochemical and functional effects of prenatal and postnatal omega-3 fatty acid deficiency on retina and brain in rhesus monkeys. Proc. Natl. Acad. Sci. U.S.A. 83, 4021–4025.

Neuringer, M., Anderson, G.J. and Connor, W.E. (1988) The essentiality of n-3 fatty acids for development and function of retina and brain. Annu. Rev. Nutr. 8, 517–541.

O'Brien, J.S. (1986) Stability of the myelin membrane. Science 147, 1099–1107.

O'Brien, J.S., Fillerup, D.L. and Mead, J.F. (1964) Quantification and fatty acid and fatty aldehyde composition of ethanolamine, choline, and serine glycerophosphatides in human cerebral gray and white matter. J. Lipid Res. 5, 329–338.

Odutuga, A.A. (1981) Reversal of brain essential fatty acid deficiency in the rat by dietary linoleate, linolenate and arachidonate. Int. J. Biochem. 13, 1035–1038.

Olendorf, W.H. (1975) Permeability of the blood–brain barrier. In: D.B. Tower (Ed.), The Nervous System, Raven Press, New York, pp. 279–289.

Op den Kamp, J.A.F., Roelofsen, B. and Van Deenen, L.L.M. (1985) Structural and dynamic aspects of phosphatidylcholine in human erythrocyte membrane. Trends Biochem. Sci. 10, 320–323.

Philbrick, D.J., Mahadevappa, V.G., Ackman, R.G. and Holub, B.J. (1987) Ingestion of fish oil or a derived n-3 fatty acid concentrate containing eicosapentaenoic acid affects fatty acid compositions of individual phospholipids of rat brain, sciatic nerve, and retina. J. Nutr. 17, 1663–1670.

Pitikin, R.M., Connor, W.E. and Lin, D.S. (1972) Cholesterol metabolism and placental transfer in pregnant rhesus monkey. J. Clin. Invest. 15, 2582–2592.

Portman, O.W., Alexander, M. and Maruffo, C.A. (1967) Nutritional control of arterial lipid composition in squirrel monkeys: major ester classes and types of phospholipids. J. Nutr. 91, 35–46.

Sanders, T.A.B., Mistry, M. and Naismith, D.J. (1984) The influence of a maternal diet rich in linoleic acid on brain and retinal docosahexaenoic acid in the rat. Br. J. Nutr. 51, 57–66.

Tarozzi, G., Barzanti, V., Biagi, P.L., Coochi, M., Lodi, R., Maranesi, M., Pignatti, C. and Turchetto, E. (1984) Fatty acid composition of single brain structures following different alpha-linolenate dietary supplementations. Acta Vitamonol. Enzymol. 6, 157–163.

Trapp, B.D. and Bernson, J. (1977) Changes in phosphoglyceride fatty acids of rat brain induced by linoleic and linolenic acids after pre and postnatal fat deprivation. J. Neurochem. 28, 1009–1013.

Walker, B.L. (1967) Maternal diet and brain fatty acids in young rats. Lipids 2, 497–500.

Yerram, N.R., Moore, S.A. and Spector, A.A. (1989) Eicosapentaenoic acid metabolism in brain microvessel endothelium: effect on prostaglandin formation. J. Lipid Res. 30, 1747–1757.

Youyou, A., Durand, G., Pascal, G., Piccotti, M., Dumont, O. and Bourre, J.M. (1986) Recovery of altered fatty acid composition induced by a diet devoid of n-3 fatty acids in myelin, synaptosomes, mitochondria and microsomes of developing rat brain. J. Neurochem. 46, 224–227.

E.R. Skinner (Ed.), *Brain Lipids and Disorders in Biological Psychiatry*

CHAPTER 3

The lipid hypothesis of schizophrenia

David Horrobin

Laxdale Ltd, Kings Park House, Laurelhill Business Park, Stirling, FK7 9JQ, Scotland,
Tel: +44 (0)1786 476000 Fax:+44 (0)1786 473137 Email: agreen@laxdale.co.uk

1. Introduction

Research in schizophrenia has been dominated by receptor-based hypotheses. Undoubtedly the most important of these has been the concept that there is excess dopaminergic function. Concepts based upon abnormalities in serotoninergic systems and upon deficits in glutamate systems have also been important.

However, all these hypotheses have been primarily driven by accidental observations of the psychotropic actions of drugs. In each case, by far the strongest evidence for each hypothesis is that drug series X blocks dopamine receptors and alleviates positive symptoms of schizophrenia, drug series Y modulates serotonin receptors and may have actions on positive and negative features of schizophrenia, or drug series Z blocks glutamate receptors and leads to some of the symptoms of schizophrenia in otherwise normal persons. There is a great shortage of evidence which is truly independent of drug action.

As a result of this we have been going round in circles for over forty years. Drugs which accidentally have been found to exert effects clinically have been put into animals and a range of psycho-pharmacological effects have been noted. These effects have then become animal models for psychotropic drug discovery. Since the models are originally based on drug action and not on any fundamental understanding of the biological basis of schizophrenia, all that they can do is guide the development of more of the same. With any luck, compounds may emerge that have greater potency or fewer side effects. But it is highly unlikely that such drug development programs will lead to the emergence of genuinely new therapeutic approaches.

I exaggerate for effect, but not much. Kendell, in his Opening Address to the 1998 CINP Congress in Glasgow, pointed out that, in terms of efficacy, the new generation of atypical neuroleptic drugs is no more effective than the typical anti-psychotics which were available forty years ago. We have exchanged one set of side effects for another which in the long term may produce more or fewer problems for patients, but that is really all we have done. Studies which look at the intention to treat outcome of cohorts of patients treated with any available antipsychotic, typical or atypical, usually reveal overall improvements in symptoms in the 15–25% range. Individual patients may do much better but they are counterbalanced by the many who fail to respond partially or completely and who drop out because of lack of efficacy, dysphoria or side effects. What that means is that in any real, unselected group of schizophrenic patients, 75–85% of the symptoms remain unaddressed.

We must do better but current drug discovery approaches will not enable us to do so. Entirely novel concepts are required which will allow an attack to be made on this huge volume of unresolved pathology. Whether it proves right or wrong, the phospholipid concept does represent such a new approach. Attempts to support or refute it are thus likely to lead to progress, irrespective of whether the concept is or is not correct.

2. *The phospholipid concept*

The phospholipid structure of the brain is extraordinarily complex. Each phospholipid molecule has a 3-carbon backbone to which are attached two fatty acids, one at an outer carbon atom (technically known as Sn-1) and the other at the middle carbon atom (known as Sn-2). At the third carbon atom (Sn-3) is attached a phosphorus atom which in turn is attached to a hydrophilic head group, usually choline, ethanolamine, serine or inositol. Since about thirty different fatty acids may be found attached to the Sn-2 and Sn-1 positions, it is clear that the variety of phospholipid molecules rivals the variety of proteins. As with proteins, apparently small differences in structure may be associated with surprisingly large changes in function.

Phospholipids have two main functions. First they comprise the basic structure of which all external and internal cell membranes are made. The proteins are often embedded in, or attached to the phospholipid membranes. Many proteins have highly hydrophobic transmembrane domains. This applies, for example, to the family of receptors which includes receptors for acetyl choline, dopamine, glutamate, serotonin and noradrenaline (Arias 1998). Because of this association with membranes, the quaternary structure and therefore function of all these proteins is determined by the structure of the immediate phospholipid environment. Small changes in the lipid environment can, for example, produce large changes in the functions of receptors and ion channels (Arias 1998, Witt and Nielsen 1994, Kang and Leaf 1996).

The second major function of phospholipids is to provide the basis for the majority of cell signalling systems. Receptor activation may produce a change in membrane-associated G-protein configuration followed by a cascade of subsequent events including activation of phospholipases, regulation of phosphorylation by protein kinases, and modulation of cyclic nucleotide metabolism and calcium movements. Phospholipids, the fatty acids released from phospholipids, and the many prostaglandins and other eicosanoids derived from these fatty acids, are involved in almost every cell-signalling system. Changes in the metabolism of membrane phospholipids are likely to have knock-on effects on almost every receptor, ion channel and other cell-signalling mechanism (Horrobin 1998a, Horrobin et al. 1994).

Of the various components of phospholipid molecules, the fatty acid attached to the Sn-2 position has a role of particular importance, both in determining the biophysical properties of the phospholipid and in the cell signalling response to challenge. In neurons and glia, the fatty acid attached to the Sn-2 position is likely to be a highly unsaturated fatty acid (HUFA), particularly arachidonic acid (AA) or docosahexaenoic acid (DHA), but also possibly dihomogammalinolenic acid (DGLA), eicosapentaenoic acid (EPA), docosapentaenoic acid (DPA) or others.

The basic proposition of the latest version of the phospholipid hypothesis is that there are in schizophrenic patients at least two different abnormalities of phospholipid metabolism relating to the Sn-2 position. The first is increased activity of one or more of the phospholipase A_2 (PLA_2) group of enzymes which is able to remove HUFAs from the Sn-2 position. The second is reduced activity of one or more components of the system which incorporates HUFAs into the Sn-2 position. Neither abnormality alone will produce schizophrenia: the simultaneous presence of both together will do so (Horrobin and Bennett 1998).

3. The main evidence

This section aims to summarise the main evidence for the phospholipid concept. The subject has been reviewed in detail several times recently (Horrobin et al. 1994, Horrobin 1998a, Horrobin and Bennett 1998).

In its most generalised form, the phospholipid hypothesis suggests that one or more of the enzymes involved in phospholipid metabolism is abnormal (Horrobin et al. 1994, Horrobin 1998b). That abnormality will lead to changes in membrane structure which will alter, to varying degrees, the quaternary structure and therefore function of all membrane-associated proteins. It will also lead to changes in the availability of cell signalling molecules. As a result there will be secondary changes in the behavior of most and perhaps all neurotransmitter related mechanisms. At present most therapeutic strategies in schizophrenia target receptors, ion channels and neurotransmitter metabolism. According to the phospholipid concept, the undoubted abnormalities in these systems which are targeted by drugs are secondary and not primary: these therapeutic approaches will therefore not deal with the primary problem, which lies in the membrane. This may well explain their relatively modest efficacy.

The assumption is that the schizophrenia phenotype may be produced by a number of different genotypes. Phospholipid abnormalities will be involved in the causation of abnormalities in many but not all patients with schizophrenia.

3.1. Elevated phospholipase A_2 activity in blood and brain

Gattaz was the first to report that there was an elevation of functionally active PLA_2 in patients with schizophrenia in plasma and platelets (Gattaz et al. 1987, 1990, 1995). Using a different assay, some have replicated this finding but others have failed (Noponen et al. 1993, Albers et al. 1993). However, recently Ross et al. (1997) using both assays were able to confirm both the positive and negative findings. There is in blood elevated PLA_2 activity of a calcium-independent form. Equally in the cortex from schizophrenic patients, calcium-independent PLA_2 activity is elevated, whereas calcium-stimulated PLA_2 activity is reduced (Ross 1999). Thus, in blood, platelets and brain there is elevation of the activity of a specific type of calcium-independent PLA_2. In animals levels of calcium-independent PLA_2 can be elevated by oxidative stress (Kuo et al. 1995). The picture has recently been complicated by the identification of a specific calcium-dependent cytosolic PLA_2 in human red cells. Concentrations of the protein are highly significantly elevated in schizophrenic patients. About half of the patients with schizophrenia have concentrations which are more than two standard deviations above the normal mean (MacDonald et al. 2000).

Two studies have reported abnormalities in schizophrenic patients in the number of poly-A repeats (Hudson et al. 1996) and the BAN-1 dimorphic site in the vicinity of the gene for PLA_2 (Ramchand et al. 1999) on chromosome 1. As is common in psychiatry, in unpublished studies one group has failed to replicate the poly-A finding, but another group has succeeded.

Animal studies have demonstrated that local application of PLA_2 into the brain produces abnormalities in function, including changes in the dopamine system, showing that a primary change in phospholipid metabolism can lead to secondary changes in neurotransmitters (Gattaz and Brunner 1996).

3.2. Abnormal phospholipid composition in red cells and brain

There have now been multiple reports which indicate that in a proportion of patients with schizophrenia there are reduced levels of highly unsaturated fatty acids, particularly AA and DHA, in red cell phospholipids (Glen et al. 1994, Peet et al. 1994, Vaddadi et al. 1996). These low levels of AA and DHA are particularly associated with negative symptoms, with tardive dyskinesia and with risk of relapse. They are not attributable to medication.

Since the AA and DHA which are low in schizophrenia are usually localised at the Sn-2 position of PL, their loss may indicate increased activity of one of the PLA_2 group of enzymes, presumably a calcium-independent form as described by Ross et al. (1997).

Although brain studies are fraught with problems, there are deficits of essential fatty acids (EFAs) in phospholipids from brains from drug-treated schizophrenic patients as compared to normal controls (Horrobin et al. 1991).

3.3. Incorporation of arachidonic acid (AA) into phospholipids

When labelled AA is added to actively metabolising cells, it is linked to coenzyme A by one of the fatty acid CoA ligase (FACL) group of enzymes, and then incorporated into PL by one of the acyl-transferase groups. Two groups of investigators have reported reduced incorporation which could be due to impaired activity of one of these types of enzymes, or to a lack of available lysophospholipid acceptor (Yao et al. 1996, Demisch et al. 1987). However, since lyso-PL levels appear to be elevated in schizophrenia, the former explanation seems more likely.

3.4. Brain phospholipid metabolism as indicated by magnetic resonance imaging

Although interpretation is not easy, ^{31}P MRI can be used as a guide to brain PL metabolism. Several groups have now reported changes in signals which can reasonably be interpreted as evidence of increased PL breakdown in both medicated and drug naïve patients (Pettegrew et al. 1991, Williamson et al. 1996). There is controversy concerning the precise interpretation of the abnormal signals but there seems little doubt that metabolism is disturbed.

3.5. Excess oxidative stress

There is now substantial evidence from a range of sources, that there is increased formation in schizophrenia of oxidised breakdown product of EFAs. This phenomenon occurs during the first untreated episode and so is not secondary to drug treatment or chronic course (Mahadik and Mukherjee 1996). Elevated oxidation can occur when EFAs are incorporated into membrane phospholipids at a reduced rate, so increasing the concentration of the fatty acids in the free, unprotected form (Cane et al. 1998). As mentioned above, increased oxidation can also activate calcium-independent PLA_2.

3.6. Abnormal neurodevelopment

There is substantial evidence from neuroanatomy, from the impact of pregnancy and perinatal factors, from the occurrence in schizophrenic patients of minor physical abnormalities, and from premorbid personality and development deviations, of failure of normal neurodevelopment in schizophrenia. The main problem is that there is no unifying biochemical concept which might explain how such failures of neurodevelopment may occur (Horrobin 1998a). Since phospholipid metabolizing enzymes play crucial roles in neuronal and synaptic growth and remodelling, it is plausible to suggest that a failure in neurodevelopment could well result from defects in fatty acid signalling.

3.7. Clinical observations on pain, inflammation and infection

Schizophrenic patients consistently show reduced pain sensation and clinical experience suggests that the impaired pain sensation is more evident during exacerbations of the disease (Horrobin et al. 1978). Schizophrenic patients in most but not all studies have been found to be at reduced risk of rheumatoid arthritis and related inflammatory disorders (Vinogradov et al. 1991). A recent large survey of mortality looked at causes of death in 10 260 schizophrenic patients as compared to the rest of the population (Brown 1997). Schizophrenic patients had increased death rates from most disorders. However, two striking observations were more than double the expected death rate for respiratory disorders (mostly infectious) and a 20% reduction in death rate from cerebrovascular disease.

All of these observations are logically explicable on the basis of reduced availability of AA in membrane PL for cell signalling purposes. AA release plays an important role in pain, and also in immune and inflammatory responses. Reduced availability of AA can explain reduced pain, reduced risk of inflammatory disorders and increased risk of death from pulmonary infections. The reduced risk of cerebrovascular disease may also be explicable in view of the important role of arachidonate metabolism in platelet function and thrombosis.

3.8. Impaired niacin flushing: defective AA-related cell signalling

A particular clinical observation which is emerging as of great value as a probe of impaired cell signalling in schizophrenia, is the niacin flushing test (Horrobin 1998b).

Niacin in high doses, as used for cholesterol lowering, very commonly produces a violent flushing reaction over the upper body. This has been shown to be due to increased formation of prostaglandin (PG) D_2 from arachidonic acid in the skin (Morrow et al. 1992, 1994). In some schizophrenic patients, the response to niacin seemed to be impaired. This could not be confirmed by investigators who used relatively low doses of niacin (Wilson and Douglass 1986, Fiedler et al. 1986), but was found by those who used a dose of 200 mg (Rybakowski and Weterle 1991). A definitive study by Hudson found flushing in 100% of 28 normal controls, in 94% of 18 bipolar patients and only 57% of 28 DSM–IV schizophrenic patients following 200 mg oral niacin (Hudson et al. 1997). There were no medication differences between flushing and non-flushing individuals. The bipolar patients flushed particularly strongly, yet 9/18 of them were on full doses of neuroleptics.

In 126 patients with predominantly negative schizophrenic symptoms 60% failed to flush to 200 mg niacin orally (Glen et al. 1996). Those who failed to flush had significantly lower red cell AA and DHA levels then those who flushed normally.

The oral niacin test can be unpleasant, is not quantitative and is time consuming. Ward et al. (1998b) therefore developed the concept of topical niacin testing in which four different concentrations of methyl nicotinate are applied to the forearm for five minutes. Over a 5–20 minute period, an adequate dose of niacin produces a strong flushing reaction confined to the area of the applied gauze pad. The color of this patch can be rated on a 0–3 scale at 5, 10 and 15 minutes using a scale standardized for background skin color. Using this test around 75% of schizophrenic patients show a substantially impaired response. In contrast, bipolar patients show a response at the upper end of the normal range. Bipolar patients show a significantly higher response to the lowest dose of methyl nicotinate (Horrobin 1998b, Ward et al. 1998a).

The response is unequivocally not attributable to medication. Peet et al. (1998) have recently demonstrated that, in drug naïve first episode patients, responses are even more impaired than in medicated patients.

The niacin test is easy to replicate and I am now aware of eight investigators who have obtained similar results which are not yet published. There is preliminary evidence that about 40–50% of first degree relatives of patients also show impaired flushing responses, which is what might be expected if the condition was dominantly transmitted.

There is only a limited number of enzymes which could explain this failure to flush and it is therefore anticipated that it will not be too long before the relevant enzymes are identified.

4. Candidate genes for involvement in schizophrenia

The proposition is that in schizophrenia there must be the simultaneous presence of two main abnormalities. One is an increased rate of loss of HUFAs from the Sn-2 position of phospholipids. The other is a reduced rate of incorporation of HUFAs into the Sn-2 position. These two main abnormalities may then be modulated by a range of other candidates which are involved in providing the supply of HUFAs to the brain. Even if both main genes are present, the phenotype may not be expressed in 50–70% of patients as indicated by concordance rates in identical twin studies.

4.1. Candidates for involvement in increased loss

The work by Gattaz et al. (1987) and by Ross et al. (1997) indicates increased activity of a calcium-independent phospholipase A_2. The recent work by MacDonald et al. (2000) suggests that there may be increased levels of a calcium-dependent cytosolic PLA_2. The abnormality may be in the PLA_2 itself or in one of the other proteins known to regulate PLA_2. These include specific G-proteins, proteins which inhibit or activate PLA_2 and receptors which inhibit PLA_2. One interesting candidate for a receptor is the $\alpha 7$ nicotinic receptor which is able to inhibit PLA_2 and which has been linked to schizophrenia (Freedman et al. 1997, Adler et al. 1998, Marin et al. 1997, Singh et al. 1998). A failure of normal $\alpha 7$ nicotinic receptor function could lead to increased PLA_2 activity.

An alternative way of releasing HUFAs from the Sn-2 position would be by the sequential action of phospholipase C to form diacylglycerol and then either PLA_2 or DAG lipase to release the HUFA from the Sn-2 position.

4.2. Candidates for involvement in reduced HUFA incorporation

There are several important candidates. First, in order to be incorporated into phospholipids, fatty acids must be linked to coenzyme A by a fatty acid-CoA ligase (FACL). FACL-4 which is found in the brain and which has high affinity for the HUFAs is one candidate (Horrobin and Bennett 2000). The HUFA–CoA must then be transported to the vacant Sn-2 position of a lysophospholipid (a phospholipid which has lost its Sn-2 fatty acid) by one of a group of acyl-transferase enzymes.

Alternatively, there are two enzyme systems which can transfer fatty acids directly from one phospholipid to another lysophospholipid without the need to form intermediate free acids or CoA derivatives. One of these systems requires coenzyme A whereas the other does not.

4.3. Candidates for modulating activities

The two primary abnormalities will increase the pool size of free HUFAs in the brain. Free HUFAs are more easily oxidised than HUFAs esterified to phospholipids or other molecules. Any of the enzymes involved in oxidative defence or in oxidation may therefore modulate the impact of the main defects.

HUFAs can be made in the brain in small amounts but most of the HUFAs used by the brain probably arrive by the blood from either the diet or the liver. Liver HUFA metabolising enzymes such as elongases and desaturases may thus have a role to play. However, the main proteins involved in transport are two fold:

(a) Albumin, which transports free fatty acids. These fatty acids may be released from albumin in the cerebral circulation by the creation of concentration gradients and as a result of several mechanisms whereby free fatty acids are removed from the cytoplasmic pool. One relates to the high activity of FACLs which quickly esterify the free acids to coenzyme A. The other relates to brain fatty acid binding proteins (FABPs) at least one of which binds HUFAs with high affinity. This FABP is strongly expressed whenever brain growth and remodelling are taking place.

(b) Lipoproteins which transport fatty acids in esterified forms in the blood. The fatty acids must be released by lipoprotein lipase which is present in the endothelium. Brain blood vessels, particularly those in the hippocampus, contain substantial amounts of lipoprotein lipase (LPL). The gene for LPL is at chromosome 8p22 which has been reorganised as one of the "hot spots" by linkage studies (Kendler et al. 1996, Kunugi et al. 1996).

4.4. Comment on candidate genes

Although there are many proteins involved directly or indirectly in phospholipid metabolism, the number of strong candidates is relatively limited. The identification of the key proteins involved is a realistic subject for a research programme. The models used on the basis of the present proposals must allow for the simultaneous presence of two genes of major effect which are expressed phenotypically only about 30–50% of the occasions they occur together. Each must also be present in 5–10% of the non-psychotic population and is likely to be influenced by a number of genes of minor effect.

5. P50 gating and niacin flushing

This section presents a hypothesis which has been developed on the basis of studies by Waldo at the University of Colorado. The Colorado research group have made a particular study of the relationship between P50 gating, the $\alpha7$ nicotinic receptor and the genetics of schizophrenia. P50 gating relates to the auditory cortex EEG response to two clicks presented in rapid succession. In normal individuals the amplitude of the EEG response to the second click is sharply reduced, usually to about 20% of the amplitude of the response to the first click. In between 70% and 90% of schizophrenic individuals, in contrast, the response to the second click is of similar amplitude to that of the first. About 7% of the normal general population show a "schizophrenic-type" P50 response. The presence of a P50 abnormality alone is therefore not diagnostic of schizophrenia.

In most families where a schizophrenic proband is the offspring of two non-psychotic parents, the child and one of the parents will manifest impaired P50 gating. Often that parent has a history of a parent, grandparent or great-grandparent who was schizophrenic. The other parent, without a P50 abnormality usually does not have any family history of schizophrenia. But that parent must have contributed something to that child to make the child schizophrenic.

Waldo wondered whether abnormal niacin flushing might be related to the P50 abnormality or, alternatively might be related to the unknown contribution from the other parent. She therefore investigated niacin flushing in the parents of ten schizophrenic children, where one parent had a P50 abnormality (Waldo 1999). All ten parents with the P50 abnormality showed clear niacin flushing responses, demonstrating that abnormal P50 habituation was not related to niacin flushing. On the other hand, of ten parents without a P50 abnormality, eight showed abnormal niacin flushing. Thus, in a high proportion of parents, abnormal niacin flushing seems to be a marker of a second abnormality, which, when linked to a P50 problem, generates a schizophrenia genotype and phenotype.

Because of the linkage between the $\alpha 7$ nicotinic receptor abnormalities and abnormal P50 gating, and because abnormal $\alpha 7$ receptor function may be linked to excess PLA_2 activity, we have proposed that abnormal P50 gating may be a marker of the excess PLA_2 half of the schizophrenia equation. If this is so, then impaired niacin flushing may be a marker of impaired incorporation of HUFAs into the Sn-2 position. Neither abnormality alone will lead to schizophrenia but the simultaneous presence of both puts the individual at risk.

It is not proposed that all schizophrenia is explicable on this basis. There will undoubtedly be other routes to schizophrenia genotypes and phenotypes. Nevertheless, the epidemiology is intriguing. Most estimates of the lifetime risk of developing schizophrenia fall in the 0.4% to 1.2% range. 9.5% of the normal population fail to flush with niacin. 7% of the normal population exhibit abnormal P50 gating. Non-linked random mating would mean that 0.7% of the population would have both abnormalities.

This is clearly a hypothesis, but a hypothesis which lends itself to testing and exploration. Both P50 gating and niacin flushing are simple non-invasive tests which can be used in patients, their relatives and in those who may be at risk of developing schizophrenia.

6. Therapeutic implications

From the point of view of the patient and society, the main criticism of most existing theories of schizophrenia is that they have been therapeutically sterile. No major improvements in outcome have resulted from the new generation of drugs, with the exception of clozapine. The best that can be said is that one set of side effects has been replaced by another. However, almost all patients remain substantially ill even when on the best therapy, and more than half of all patients drop out of full compliance with treatment programmes. Something more is needed and concepts of pathogenesis need to be judged on the basis of whether or not they are therapeutically fruitful.

Optimal therapeutic success will result from identification of which proteins are abnormal and then correcting their function. However, this strategy may in practice be more difficult than in theory. It is forty years since the abnormal structure of sickle cell haemoglobin was identified, yet this has not led to much therapeutic progress in the disease. Similarly, therapeutic progress has not followed the identification of the genes and their protein products responsible for cystic fibrosis and muscular dystrophy. An alternative strategy may be to look at what these enzyme systems do and use that knowledge to manipulate substrate–product relationships and cell signalling mechanisms in an appropriate way.

One of the most obvious things to do might be to attempt to correct the composition of the PL at the Sn-2 position by providing the brain-related EFAs, and perhaps reducing intake of saturated fats which will tend to displace them and make any abnormality worse. An indication that this strategy might be fruitful came from the WHO study on the epidemiology of schizophrenia in eight countries (World Health Organization 1979).

While the lifetime incidence of schizophrenia was virtually identical in all eight, there were large differences in severity. In some countries schizophrenia tended to be a severely disabling disorder, while in others it tended to have a more benign course with frequent prolonged remissions.

The explanation for this variation in severity is unknown and many unsuccessful attempts have been made to explain it on the basis of social, family or employment structures, or of medical care and economic systems. All of these factors may make some contribution but the disturbing fact is that outcome is consistently poorer in those countries with advanced medical care systems. Christensen and Christensen (1988), in contrast, pointed out that diet might be a major factor. Saturated fat intake is very high especially in relation to fish and vegetable fat intake, in countries with a poor outcome. Correlations between the ratio of animal fat to vegetable and fish fat suggest that these could explain 90% of the variability in outcome of the disease (Christensen and Christensen 1988, Horrobin 1992).

If this is so, then treatment with brain related EFAs might be beneficial. The structural EFA, DHA, the structural and cell-signalling AA, and the cell signalling EPA and DGLA (or its precursor GLA) are all candidates. Preliminary studies with GLA showed only modest beneficial effects (Vaddadi 1992) whereas AA appeared to worsen symptoms in some patients. If AA cannot be incorporated normally into membrane PL it may be a waste of time using it. Fish oil administration, which contains both DHA and EPA appeared to be beneficial but could not distinguish between EPA and DHA (Mellor et al. 1996).

Peet et al. decided to compared a corn oil placebo with EPA and DHA given separately in a randomised study in chronic treatment-resistant patients (Peet et al. 1997, Shah et al. 1998). The expectation was that DHA, as the most abundant brain fatty acid, would be the most beneficial. The EFAs were added to existing medications. The outcome was a surprise. DHA essentially had no effect, placebo produced a modest improvement, but EPA had an effect significantly better than DHA or placebo, and one which was comparable in magnitude to that produced by the new generation of atypical antipsychotics.

This positive effect of EPA has been confirmed in two open studies (Shah et al. 1998, Puri and Richardson 1998), and further randomised, placebo-controlled trials are in progress. One of the open studies, a case report, was important because EPA (in the form of Kirunal, an EPA-enriched fish oil) was used as monotherapy in a drug-naïve patient who had been ill for several years but refused all treatment (Puri and Richardson 1998). This patient was essentially normalised by EPA alone, returned to university and is close to completing a degree (Puri et al. 2000).

Very recently two further randomised, placebo-controlled trials have been completed using EPA. Peet and Shah randomised 30 new episode unmedicated patients to receive an EPA-enriched fish oil (Kirunal) or matching placebo. The psychiatrist responsible for the patients then prescribed standard antipsychotic drugs as required. After 12 weeks the Kirunal group had improved substantially more than the placebo group. More importantly, whereas all of the placebo group required full doses of standard antipsychotic drugs, 43% of the EPA group did not require any antipsychotic therapy other than Kirunal (Shah et al. 2000). Very preliminary results have also been presented of a study of pure

EPA (LAX–101) in chronic, partially treatment-unresponsive schizophrenic patients (Peet and Horrobin 2000). This study also demonstrated that EPA was effective in treating schizophrenic symptoms without producing any of the side effects characteristic of standard old or new antipsychotic drugs.

A mixture of EPA and DHA has also been found to be beneficial in bipolar disorder in a randomised, placebo-controlled study (Stoll et al. 1999). Several drugs are useful in both bipolar disorder and schizophrenia and so there is nothing surprising about the fact that one compound may work for both.

EPA is probably acting as a second messenger/cell signalling molecule rather than a structural element. Its impact on outcome needs to be explored in a wide range of studies. The precise target remains uncertain. One possible candidate is FACL-4, the brain specific acyl-CoA ligase (Horrobin and Bennett 2000). This enzyme is unique in that it has an affinity for EPA which is several fold higher than that for AA or DHA (Cao et al. 1998). The human enzyme is located on chromosome X and might possibly therefore explain some of the sex differences in schizophrenia.

7. Conclusions

There is substantial evidence for the PL hypothesis which appears to explain a higher proportion of the known facts about schizophrenia than other concepts. The big challenge remains the identification of the specific candidate proteins and the translation of the initial tantalising therapeutic observations into large scale, thoroughly replicated results. Because of the safety of EPA, and the possibility of identifying adolescents at high risk by P50 gating and niacin flushing studies, trials of EPA as a preventive agent should also be considered.

References

Adler, L.E., Olincy, A., Waldo, M., Harris, J.G., Griffith, J., Stevens, K., Flach, K., Nagamoto, H., Bickford, P., Leonard, S. and Freedman, R. (1998) Schizophrenia, sensory gating, and nicotinic receptors. Schizophr. Bull. 24, 189–202.

Albers, M., Meurer, H., Marki, F. and Klotz, J. (1993) Phospholipase A2 activity in the serum of neuroleptic-naive psychiatric inpatients. Pharmacopsychiatry 26, 94–98.

Arias, H.R. (1998) Binding sites for exogenous and endogenous non-competitive inhibitors of the nicotinic acetylcholine receptor. Biochim. Biophys. Acta Rev. Biomembranes 1376, 173–220.

Brown, S. (1997) Excess mortality of schizophrenia. A meta-analysis. Br. J. Psychiatry 171, 502–508.

Cane, A., Breton, M., Koumanov, K., Bereziat, G. and Colard, O. (1998) Impairment of fatty acid acylation accounts for oxidant-induced arachidonic acid release in vascular smooth muscle cells. In: Abstracts 3rd ISSFAL Congress, Lyon, France, June 1–5, Abstract 166.

Cao, Y., Traer, E., Zimmerman, G.A., McIntyre, T.M. and Prescott, S.M. (1998) Cloning, expression, and chromosomal localization of human long-chain fatty acid CoA ligase 4 (FACL4). Genomics 49, 327–330.

Christensen, O. and Christensen, E. (1988) Fat consumption and schizophrenia. Acta Psychiatr. Scand. 78, 587–591.

Demisch, L., Gerbaldo, H., Gebhart, P., Georgi, K. and Bochnik, H.J. (1987) Incorporation of ^{14}C-arachidonic acid into platelet phospholipids of untreated patients with schizophreniform or schizophrenic disorders. Psychiatr. Res. 22, 275–282.

Fiedler, P., Wolkin, A. and Rotrosen, J. (1986) Niacin-induced flush as a measure of prostaglandin activity in alcoholics and schizophrenics. Biol. Psychiatry 21, 1347–1350.

Freedman, R., Coon, H., Worsley, M.M., Urtreger, A.O., Olincy, A., Davis, A., Polymeropoulos, M., Holik, J., Hopkins, J., Hoff, M., Rosenthal, J., Waldo, M.C., Reimherr, F., Wender, P., Yaw, J., Young, D.A., Breese, C.R., Adams, C., Patterson, D., Adler, L.E., Kruglyak, L., Leonard, S. and Byerley, W. (1997) Linkage of a neurophysiological deficit in schizophrenia to a chromosome 15 locus. Proc. Natl. Acad. Sci. U.S.A. 94, 587–92.

Gattaz, W.F. and Brunner, J. (1996) Phospholipase A2 and the hypofrontality hypothesis of schizophrenia. Prostaglandins Leukotrienes Essent. Fatty Acids 55, 109–113.

Gattaz, W.F., Kollisch, M., Thurn, T., Virtanen, J.A. and Kinnunen, P.K.J. (1987) Increased plasma phospholipase-A2 activity in schizophrenic patients: reduction after neuroleptic therapy. Biol. Psychiatry 22, 421–426.

Gattaz, W.F., Hubner, C.V.K. and Nevalainen, T.J. (1990) Increased serum phospholipase A2 activity in schizophrenia: a replication study. Biol. Psychiatry 28, 495–501.

Gattaz, W.F., Schmitt, A. and Maras, A. (1995) Increased platelet phospholipase A2 activity in schizophrenia. Schizophr. Res. 16, 1–6.

Glen, A.I.M., Glen, E.M., Horrobin, D.F., Vaddadi, K.S., Spellman, M., Morse-Fisher, N., Ellis, K. and Skinner, F.S. (1994) A red cell membrane abnormality in a subgroup of schizophrenic patients: evidence for two diseases. Schizophr. Res. 12, 53–61.

Glen, A.I.M., Cooper, S.J., Rybakowski, J., Vaddadi, K., Brayshaw, N. and Horrobin, D.F. (1996) Membrane fatty acids, niacin flushing and clinical parameters. Prostaglandins Leukotrienes Essent. Fatty Acids 15, 9–15.

Horrobin, D.F. (1992) The relationship between schizophrenia and essential fatty acid and eicosanoid metabolism. Prostaglandins Leukotrienes Essent. Fatty Acids 46, 71–77.

Horrobin, D.F. (1998a) The membrane phospholipid hypothesis as a biochemical basis for the neurodevelopmental concept of schizophrenia. Schizophr. Res. 30, 193–208.

Horrobin, D.F. (1998b) Niacin cutaneous flushing as a probe of fatty acid related cell-signalling mechanisms in schizophrenia and affective disorders. Presented at NIH Workshop on Omega-3 Essential Fatty Acids and Psychiatric Disorders, Bethesda, MD, 2–3 September 1998.

Horrobin, D.F. and Bennett, C.N. (1998) The membrane phospholipid concept of schizophrenia. Presented at IVth Symp. on the Search for the Causes of Schizophrenia, Guaruja, Brazil, 10–13 November 1998.

Horrobin, D.F. and Bennett, C.N. (2000) Fatty acid coenzyme A ligases as candidate genes in schizophrenia. Schizophr. Res. 41, 99.

Horrobin, D.F., Ally, A.I., Karmali, R.A., Karmazyn, M., Manku, M.S. and Morgan, R.O. (1978) Prostaglandins and schizophrenia: further discussion of the evidence. Psychol. Med. 8, 43–48.

Horrobin, D.F., Manku, M.S., Hillman, H., Iain, A. and Glen, M. (1991) Fatty acid levels in the brains of schizophrenics and normal controls. Biol. Psychiatry 30, 795–805.

Horrobin, D.F., Glen, A.I. and Vaddadi, K. (1994) The membrane hypothesis of schizophrenia. Schizophr. Res. 13, 195–207.

Hudson, C.J., Lin, A. and Horrobin, D.F. (1996) Phospholipases: in search of a genetic base of schizophrenia. Prostaglandins Leukotrienes Essent. Fatty Acids 55, 119–122.

Hudson, C.J., Lin, A., Cogan, S., Cashman, F. and Warsh, J.J. (1997) The niacin challenge test: Clinical manifestation of altered transmembrane signal transduction in schizophrenia? Biol. Psychiatry 41, 507–513.

Kang, J.X. and Leaf, A. (1996) Evidence that free polyunsaturated fatty acids modify Na^+ channels by directly binding to the channel proteins. Proc. Natl. Acad. Sci. U.S.A. 93, 3542–3546.

Kendler, K.S., Maclean, C.J., O'Neill, A., Burke, J., Murphy, B., Duke, F., Shinkwin, R., Easter, S.M., Webb, B.T., Zhang, J., Walsh, D. and Straub, R.E. (1996) Evidence for a schizophrenia vulnerability locus on chromosome 8p in the Irish study of high-density schizophrenia families. Am. J. Psychiatry 153, 134–154.

Kunugi, H., Curtis, D., Vallada, H.P., Nanko, S., Kunugi, H., Powell, J.F., Murray, R.M., McGuffin, P., Owen, M.J., Gill, M. and Collier, D.A. (1996) A linkage study of schizophrenia with DNA markers from chromosome 8p21-p22 in 25 multiplex families. Schizophr. Res. 22, 61–68.

Kuo, C.F., Cheng, S. and Burgess, J.R. (1995) Deficiency of vitamin E and selenium enhances calcium-independent phospholipase A2 activity in rat lung and liver. J. Nutr. 125, 1419–1429.

MacDonald, D.J., Glen, A.C.A., Boyle, R.M., Glen, A.I.M., Ward, P. and Horrobin, D.F. (2000) An Elisa method for type IV cPLA2 in RBC: a potential marker for schizophrenia. Schizophr. Res. 41, 259.

Mahadik, S.P. and Mukherjee, S. (1996) Free radical pathology and antioxidant defense in schizophrenia: a review. Schizophr. Res. 19, 1–17.

Marin, P., Hamon, B., Glowinski, J. and Premont, J. (1997) Nicotine-induced inhibition of neuronal phospholipase A2. J. Pharmacol. Exp. Ther. 280, 1277–1283.

Mellor, J.E., Laugharne, J.D.E. and Peet, M. (1996) Omega-3 fatty acid supplementation in schizophrenic patients. Hum. Psychopharmacol. 11, 39–46.

Morrow, J.D., Awad, J.A., Oates, J.A. and Roberts, L.J. (1992) Identification of skin as a major site of prostaglandin D2 release following oral administration of niacin in humans. J. Invest. Dermatol. 98, 812–815.

Morrow, J.D., Minton, T.A., Awad, J.A. and Roberts, L.J. (1994) Release of markedly increased quantities of prostaglandin D2 from the skin in vivo in humans following the application of sorbic acid. Arch. Dermatol. 130, 1408–1412.

Noponen, M., Sanfilipo, M., Samanich, K., Ryer, H., Ko, G., Angrist, B., Wolkin, A., Duncan, E. and Rotrosen, J. (1993) Elevated PLA-2 activity in schizophrenics and other psychiatric patients. Biol. Psychiatry 34, 641–649.

Peet, M. and Horrobin, D.F. (2000) A multicentre trial of ethyl eicosapentaenoate in schizophrenia. Schizophr. Res. 41, 225.

Peet, M., Laugharne, J.D.E., Horrobin, D.F. and Reynolds, G.P. (1994) Arachidonic acid: a common link in the biology of schizophrenia? Arch. Gen. Psychiatry 51, 665–666.

Peet, M., Laugharne, J.D.E. and Mellor, J. (1997) Double blind trial of n-3 fatty acid supplementation in the treatment of schizophrenia. International Congress on Schizophrenia Research, Colorado Springs.

Peet, M., Shah, S. and Ramchand, C.N. (1998) Niacin challenge skin flushing in drug-treated, drug-naive schizophrenics and matched controls: A study conducted in India. Presented at XXIst CINP Congress, Glasgow, 12–16 July 1998.

Pettegrew, J.W., Keshavan, M.S., Panchalingam, K., Strychor, S., Kaplan, D.B., Tretta, M.G. and Allen, M. (1991) Alterations in brain high-energy phosphate and membrane phospholipid metabolism in first-episode, drug-naive schizophrenics. Arch. Gen. Psychiatry 48, 563–568.

Puri, B.K. and Richardson, A.J. (1998) Sustained remission of positive and negative symptoms of schizophrenia following dietary supplementation with polyunsaturated fatty acids. Arch. Gen. Psychiatry 55, 188–189.

Puri, B.K., Richardson, A.J., Horrobin, D.F., Easton, T., Saeed, N., Oatridge, A., Hajnal, J.V. and Bydder, G.M. (2000) Eicosapentaenoic acid treatment in schizophrenia associated with symptom remission, normalization of blood fatty acids, reduced neuronal membrane phospholipid turnover and structural brain changes. Int. J. Clin. Pract. 54, 57–63.

Ramchand, C.N., Wei, J., Lee, K.H. and Peet, M. (1999) Phospholipase A2 gene polymorphism and associated biochemical alterations in schizophrenia. In: M. Peet, I. Glen and D.F. Horrobin (Eds.), Phospholipid Spectrum Disorder in Psychiatry, Marius Press, Carnforth, UK, pp. 31–37.

Ross, B.M. (1999) Brain and blood phospholipase activity in psychiatric disorders. In: M. Peet, I. Glen and D.F. Horrobin (Eds.), Phospholipid Spectrum Disorder in Psychiatry, Marius Press, Carnforth, UK, pp. 23–29.

Ross, B.M., Hudson, C., Erlich, J., Warsh, J.J. and Kish, S.J. (1997) Increased phospholipid breakdown in schizophrenia – Evidence for the involvement of a calcium-independent phospholipase A(2). Arch. Gen. Psychiatry 54, 487–494.

Rybakowski, J. and Weterle, R. (1991) Niacin test in schizophrenia and affective illness. Biol. Psychiatry 29, 834–836.

Shah, S., Vankar, G.K., Telang, S.D., Ramchand, C.N. and Peet, M. (1998) Eicosapentaenoic acid (EPA) as an adjunct in the treatment of schizophrenia. Schizophr. Res. 29, 158.

Shah, S., Ramchand, C.N. and Peet, M. (2000) Are polyunsaturated fatty acids a serious innovation in treatment for schizophrenia? Schizophr. Res. 41, 27.

Singh, I.N., Sorrentino, G., Sitar, D.S. and Kanfer, J.N. (1998) (−)Nicotine inhibits the activations of phospholipases A2 and D by amyloid beta peptide. Brain Res. 800, 275–281.

Stoll, A.L., Severus, W.E., Freeman, M.P., Rueter, S., Zboyan, H.A., Diamond, E., Cress, K.K. and Marangell, L.B. (1999) Omega 3 fatty acids in bipolar disorder – A preliminary double-blind, placebo-controlled trial. Arch. Gen. Psychiatry 56, 407–412.

Vaddadi, K.S. (1992) Use of gamma-linolenic acid in the treatment of schizophrenia and tardive dyskinesia. Prostaglandins Leukotrienes Essent. Fatty Acids 46, 67–70.

Vaddadi, K.S., Gilleard, C.J., Soosai, E., Polonowita, A.K., Gibson, R.A. and Burrows, G.D. (1996) Schizophrenia, tardive dyskinesia and essential fatty acids. Schizophr. Res. 20, 287–294.

Vinogradov, S., Gottesman, I.I., Moises, H.W. and Nicol, S. (1991) Negative association between schizophrenia and rheumatoid arthritis. Schizophr. Bull. 19, 669–678.

Waldo, M.C. (1999) Co-distribution of sensory gating and impaired niacin flush response in the parents of schizophrenics. Schizophr. Res. 40, 49–53.

Ward, P.E., Glen, E., Sutherland, J. and Glen, A. (1998a) The response of the niacin skin test in the functional psychoses. Presented at XXIst CINP Congress, Glasgow, 12–16 July 1998.

Ward, P.E., Sutherland, J., Glen, E.M.T. and Glen, A.I.M. (1998b) Niacin skin flush in schizophrenia: a preliminary report. Schizophr. Res. 29, 269–274.

Williamson, P.C., Brauer, M., Leonard, S., Thompson, T. and Drost, D. (1996) ^{31}P magnetic resonance spectroscopy studies in schizophrenia. Prostaglandins Leukotrienes Essent. Fatty Acids 55, 115–118.

Wilson, D.W.S. and Douglass, A.B. (1986) Niacin skin flush is not diagnostic of schizophrenia. Biol. Psychiatry 21, 971–974.

Witt, M.R. and Nielsen, M. (1994) Characterisation of the influence of unsaturated free fatty acids on brain GABA/benzodiazepine receptor binding in vitro. J. Neurochem. 62, 1432–1439.

World Health Organization (1979) Schizophrenia: An International Follow-up Study, Wiley, New York.

Yao, J.K., van Kammen, D.P. and Gurklis, J.A. (1996) Abnormal incorporation of arachidonic acid into platelets of drug-free patients with schizophrenia. Psychiatr. Res. 60, 11–21.

E.R. Skinner (Ed.), *Brain Lipids and Disorders in Biological Psychiatry*
© 2002 Elsevier Science B.V.   All rights reserved

Apolipoprotein E and lipid mobilization in neuronal membrane remodeling and its relevance to Alzheimer's disease

Marc Danik

Douglas Hospital Research Centre, Faculty of Medicine, McGill University, Montréal, Québec, Canada

Judes Poirier

*Douglas Hospital Research Centre and McGill Centre for Studies in Aging,
Department of Psychiatry, Neurology and Neurosurgery, McGill University, Montréal, Québec, Canada*

Abbreviations

DiI-LDL	1,1'-dioctadecyl 3,3,3',3'-tetramethylindocarbocyanine-LDL
CA4	cornus ammonis sector 4
GFAP	glial fibrillary acidic protein

1. Introduction

Interest in the importance of apolipoprotein E (apoE) in the central nervous system (CNS) grew substantially since the first published reports linking one of the corresponding three common human alleles, namely the APOE ε4 allele, to both familial and sporadic late-onset Alzheimer's disease (AD) (Corder et al. 1993, Saunders et al. 1993, Poirier et al. 1993a). Accordingly, the ε4 allele frequency was shown to increase significantly (~3-fold, i.e., from 14% up to 40–50%) in the Alzheimer population (Strittmatter et al. 1993, Poirier et al. 1993a, Farrer et al. 1997). Furthermore, a gene dosage effect was observed which translates into the fact that inheritance of at least one APOE ε4 allele is associated with a dose-related higher risk and younger age of onset distribution of AD (Corder et al. 1993, Poirier et al. 1993a, Nalbantoglu et al. 1994, Farrer et al. 1997). The mechanism whereby apoE modulates the onset of the disease is still elusive, although several hypotheses have been put forward. Some of these hypotheses served as guidelines in the attempt to link apoE to either one of two of the hallmarks of Alzheimer's disease, namely amyloid plaque formation and neurofibrillary tangles. Since the discussion of these hypotheses and the associated findings are beyond the scope of this chapter, the reader is referred to the recent reviews of Higgins et al. (1997) and Beffert et al. (1998b) for further details.

Alzheimer's disease is a progressive neurodegenerative disorder and the most frequent cause of dementia. The latter is defined as a decline of intellectual function that is associated with changes in behavior and impairment of social and professional activities, and is reflected in activities of daily living. Memory loss is a major feature of the clinical syndrome. We believe that the progression of AD is strongly dependent on the functional role of apoE in lipid transport and homeostasis in the CNS, and its ability to assist in tissue repair functions. Cellular plasticity and synaptic replacement are normally observed in a healthy brain in response to synaptic loss. One can thus assume that reinnervation processes become seriously compromised in cases where there is a dysfunction or a dysregulation of the lipid homeostatic system. This system appears to be very much sollicitated in AD, as neuronal death with accompanying synaptic losses are hallmarks and the ultimate causes of the disease. Early data from animal lesion paradigms of the peripheral and central nervous systems, such as the sciatic nerve crush (Boyles et al. 1989, Goodrum 1991) and the lesioning of the entorhinal cortex (Poirier et al. 1991b, 1993b), respectively, suggested that apoE plays a role in the coordinated storage and redistribution of cholesterol and phospholipids among cells within the remodeling area. ApoE is now believed to play an important role not only in reactive synaptogenesis, by delivering lipids to remodeling and sprouting neurons in response to tissue injury, but also in physiological ongoing synaptic plasticity and maintenance of neuronal integrity, as well as in cholinergic activity (Poirier 1994, Masliah et al. 1995, Stone et al. 1997).

In this chapter, we shall start with a brief description of apolipoprotein E and its gene, and then review the experimental data that support the role of apoE as a modulator of lipid homeostasis and synaptic plasticity. The well established peripheral nerve model of axonal regeneration and remyelination involving apoE and LDL receptors will be presented first. This model has been extensively studied in regards to lipid mobilization and fate following neuronal injury, and has been used as a reference for the characterization of apoE's function in the CNS. The entorhinal cortex lesioning (ECL) model, which mimics certain neuropathological aspects of AD, will be introduced next in conjunction with the discovery that apoE plays a pivotal role in the events that follow hippocampal deafferentation. In the last section, we will discuss the relevance to AD, of the importance of apoE in lipid mobilization and delivery of precursors to neurons for the syntheses of cellular membranes and neurotransmitters such as acetylcholine.

2. Apolipoprotein E

2.1. Human alleles and isoforms

ApoE is a polypeptide that is secreted as a glycosylated single-chain of 299 amino acids (Rall et al. 1982). In humans, its apparent molecular weight, as determined by SDS–PAGE under reducing conditions, is approximately 34 kDa. It is encoded by a four-exon 3.6 kb gene located on the long arm of chromosome 19 (Das et al. 1985). Multiple alleles at a single APOE genetic locus are responsible for the three major protein isoforms found in humans (Zannis and Breslow 1981). These isoforms, designated apoE2, apoE3, and apoE4, are the products of alleles ε2, ε3, and ε4, respectively. Therefore, the latter give rise to the three most common homozygous phenotypes, which are E2/2, E3/3, and E4/4, and the three most common heterozygous phenotypes, E2/3, E2/4, and E3/4. Among APOE alleles, ε3 is the most common. Its frequency in a typical Caucasian population is approximately 78%, whereas the frequency of the ε2 and ε4 alleles are 8% and 14%, respectively (Farrer et al. 1997). Allelic distribution in other ethnic groups such as African Americans, Hispanics, and Japanese were reported to be quite similar

(Farrer et al. 1997). The allelic heterogeneity gives rise to a protein polymorphism at two positions: residues 112 and 158 on the mature protein. In comparison to apoE3, apoE2 and apoE4 contain substitutions of cysteine for arginine at residue 158 and arginine for cysteine at residue 112, respectively. These single amino acid substitutions lead to a charge difference detectable by isoelectric focusing electrophoresis (Utermann et al. 1979).

2.2. Plasma apoE

It is known from cardiovascular research that carriers of different apoE genotypes are at different risk for developing atherosclerosis (Davignon et al. 1988). The allele-specific effects on lipoprotein metabolism and profile are believed to stem from structural and physiological differences in the properties between the apoE isoforms, and their consequence on the individual's plasma apoE levels (Rubinsztein 1995). Indeed, apoE has been extensively studied in non-nervous tissues as one of several apolipoproteins that direct lipid metabolism. It is synthesized in most organs (Elshourbagy et al. 1985) and is believed to transport lipids between cells not only of different organs but also within a given tissue. In the vascular compartment, apoE is found in association with several classes of lipoproteins including chylomicrons and very low-density lipoproteins (VLDL), remnants of these two classes, and a subclass of the high-density lipoproteins (HDL) (Mahley 1988). These apoE-containing lipoprotein complexes carry lipids such as cholesterol, cholesterol esters, phospholipids, and triglycerides to be used as stores of chemical energy (primarily triglycerides) or as structural components of membranes (phospholipids and cholesterol), or, alternatively, to be disposed of into bile acids. ApoE serves as a ligand for several cell surface receptors that mediate the uptake of these lipoprotein particles (Gliemann 1998).

2.3. Brain apoE

In humans, the brain is the most important site of apoE expression next to the liver (Elshourbagy et al. 1985). ApoE was shown to be synthesized and secreted by glial cells of rodents and humans, predominantly by astrocytes (Boyles et al. 1985, Poirier et al. 1991b, Diedrich et al. 1991, Nakai et al. 1996). Intraneuronal localization of human apoE has been observed in several studies (Han et al. 1994, Schmechel et al. 1996) and is consistent with *in vitro* data showing that, in addition to astrocytes (Pitas et al. 1987), primary cultures of fetal rat hippocampal neurons demonstrate the capacity to internalize apoE-containing lipoproteins (Beffert et al. 1998a). In agreement, all known receptors for apoE have been shown to be present in the mammalian brain on one or several cell types including neurons (for a review, see Beffert et al. 1998b). These receptors are members of a single family and include the low-density lipoprotein (LDL) receptor, the very low-density lipoprotein (VLDL) receptor, the apoER2 receptor, the LDL receptor-related protein (LRP), and the gp330/megalin receptor. Furthermore, apoE was shown to be present on lipoprotein particles of the cerebrospinal fluid (CSF) (Borghini et al. 1995, LaDu et al. 1998). Together, these data suggest that apoE also participates in lipoprotein metabolism and lipid homeostasis in the CNS. The importance of apoE in the nervous tissue is underscored by the fact that major plasma apolipoproteins such as apoB (typical of LDL) and apoAI (typical of HDL) are not synthesized within the tissue and have very limited access to the brain, if not at all (Weiler-Guttler et al. 1990, Osman et al. 1995).

3. The role of apolipoprotein E in peripheral nerve repair

3.1. Wallerian degeneration and secondary neural lipid changes

In the mammalian peripheral nervous system, the interruption of axons by disease or injury leads to the initiation of a sequence of morphological events illustrated by degenerative processes and subsequent regenerative responses and, as long as the local environment is favorable, the recovery of function (for a review, see Goodrum and Bouldin 1996). Breakdown of axons causes fragmentation of surrounding myelin into blocks or ovoids in which lie fragments of axons. This is a feature typical of a process called Wallerian degeneration, in which degeneration of both axon and myelin sheaths occur distal to axonal interruption. Within the first three days following nerve injury, neighboring Schwann cells initiate the degradation of the ovoids by sequestering them and reducing them to lipid droplets within their cytoplasm. Blood-borne inflammatory cells rapidly infiltrate the distal nerve stump. Among them, the monocyte-derived macrophages significantly, along with the participation of resident macrophages, contribute to the bulk of phagocytosis and degradation of myelin debris. As axonal and myelin debris are being cleared during the first few days following injury, the axons proximal to the site of injury respond by growing back through the distal neurolemmal tube which contains proliferated Schwann cells. The Schwann cells start to remyelinate the regenerated axons within the first two weeks by using first its stored lipids then those supplied by the macrophages.

The sciatic nerve crush paradigm in the rat has been extensively used as a model to study Wallerian degeneration and subsequent regeneration processes, both at the cellular and molecular levels. Using this model, it was shown that the lipid composition and content of the degenerating nerve change as the breakdown and removal of myelin debris proceed. Total *de novo* lipid synthesis in Schwann cells decreases rapidly following a nerve crush. With the exception of cholesteryl esters, myelin lipids (cholesterol and cerebrosides in particular) synthesis is inhibited by three days as a consequence of cholesterol and fatty acid release from degenerating myelin (White et al. 1989). A minor portion of the phospholipids are hydrolyzed by phospholipases to generate free fatty acids that are either recycled and stored within the Schwann cell, as lipid droplets of cholesteryl esters and glycerolipids, or secreted and taken-up by cells (fibroblasts and macrophages) outside the nerve fiber (White et al. 1989).

Degradation of myelin lipids is slow until macrophages appear, in which much of the myelin degradation eventually occurs. Most of the myelin-specific proteins and myelin lipids, as well as other lipids, fall dramatically or disappear from the nerve by three weeks after the onset of Wallerian degeneration (Wood and Dawson 1974). Accordingly, the content of myelin-enriched lipids, such as cerebrosides and sphingomyelin, was shown to decrease by 70–80%, whereas the phospholipid content of the nerve decreased by approximately 50% (Wood and Dawson 1974). Unlike most other membrane components, myelin cholesterol appears to be retained within the nerve following injury, as total cholesterol content of the nerve did not change significantly over the entire course of degeneration and regeneration (Yao et al. 1980). Indeed there is evidence that the cholesterol from degenerating myelin is salvaged and stored within the macrophages and subsequently delivered to Schwann cells for the formation of new myelin during regeneration of the nerve (Boyles et al. 1989, Goodrum 1991). A portion of the fatty acids generated by myelin degradation in macrophages appears to have the same fate, as they are ultimately reincorporated into phospholipids in the newly synthesized myelin (Goodrum et al. 1995).

3.2. Nerve regeneration and lipid mobilization

The recycling of lipids, in particular myelin cholesterol, during peripheral nerve regeneration in the mouse was first described in the early seventies (Rawlins et al. 1970). Further characterization of the sciatic nerve lesion paradigm in the rat confirmed that some lipids were reutilized and also led to the proposition of a model for lipid mobilization following injury that implicated apoE (Mahley 1988, Goodrum and Bouldin 1996). The discovery that apoE is involved in the mechanism by which cholesterol and fatty acids from degenerating myelin move from macrophages to Schwann cells and regenerating axons, resulted from the search of newly synthesized proteins during Wallerian degeneration (Ignatius et al. 1986, Snipes et al. 1986). It was shown that apoE synthesis and secretion originated solely from the resident and infiltrated macrophages, that they markedly increased by day 3, peaked between days 7 and 10, were greatly reduced by day 28, and were almost back to constitutive levels by the fifteenth week after injury, when the nerve appeared to be essentially normal (Boyles et al. 1989).

Current evidence suggests that during the first week following a lesion to the sciatic nerve, apoE is secreted and incorporated into lipoprotein disks and spheres derived from degenerating myelin. These particles are internalized by endoneurial macrophages in order to assist phagocytosis in the salvage of lipids. The cholesterol and fatty acids from phospholipids are first stored in lipid droplets as cholesteryl esters and glycerolipids (Goodrum et al. 1995), and then subsequently secreted by the macrophages to form cholesterol-rich, apoE-containing lipoproteins, which are taken-up via low-density lipoprotein receptors in Schwann cells that have started to remyelinate regenerating axons (Boyles et al. 1989). Accordingly, the level of LDL receptors along Schwann cells increased during the third week and remained elevated until the fifteenth week. Consistent with a receptor-mediated uptake of lipoproteins by Schwann cells, cholesterol synthesis in regenerating nerve is down-regulated at the level of hydroxymethylglutaryl-CoA (HMG-CoA) reductase (Goodrum 1990).

In support of this lipid mobilization model, apoE begins to accumulate in the extracellular matrix during the second week post-lesion and can be found on the surface of Schwann cells. Peak accumulation (which is around 250-fold) between the third and fourth week coincide with increasing concentration of discoidal and small spherical lipoprotein particles in the extracellular matrix, Schwann cell depletion of their cholesteryl esters stores, and the thickening of myelin sheaths (Boyles et al. 1990). From the analysis of lipoproteins isolated from injured nerves, it was shown that apoE is indeed associated with heparin-binding particles. The protein content of these particles, which also included other protein species, represented 29% of their mass. Other components were phospholipids (33%) and cholesterol (38%) of which 77% was nonesterified, in keeping with the discoidal nature of the majority of the particles (Boyles et al. 1989).

In addition to supplying Schwann cells with lipids for remyelination, apoE is believed to deliver cholesterol and phospholipids to sprouting and elongating axons so they can be incorporated into newly and massively synthesized neuronal membranes. Axon regrowth has been observed to start within the first two days after nerve injury with concomitant high levels of LDL receptor expression on advancing tips (Boyles et al. 1989).

3.3. Other apolipoproteins

In addition to apoE, several apolipoproteins including apoAI, apoAIV and apoD (Boyles et al. 1989, 1990), as well as apoJ mRNA (Bonnard et al. 1997), have been identified in the regenerating nerve and may contribute to the intercellular movement of lipids. All of these apolipoproteins are typically found on HDL particles in body fluids and could be thus involved in the promotion of cholesterol and phospholipid efflux from lipid loaded cells. Apolipoproteins AI and AIV appear to enter the nerve from the plasma as a result of disruption of the blood–nerve-barrier following denervation. Like plasma albumin, they slowly accumulated within the nerve until the third week after injury, with relative peak concentrations of 14 to 26-fold (Boyles et al. 1990). In contrast, apoD gene expression is induced rapidly and locally in endoneurial fibroblasts, particularly in the distal segment of the crushed nerve, with maximum transcript levels (40-fold above control) observed at day 6, coinciding with the period of nerve repair (Spreyer et al. 1990). The protein was shown to accumulate in the endoneurial extracellular space up to 500-fold between the third and fourth week and was found to copurify with apoE in the lipoprotein fraction (Boyles et al. 1990, Spreyer et al. 1990). Thus, apoD and apoE were both rapidly increasing in concentrations during axon growth, and were both present at high levels during active remyelination. The function of apoD in peripheral nerve repair remains unclear. It was proposed that apoD has a role in lipoprotein formation, perhaps in relation to cholesterol esterification, or in their transport for membrane biosynthesis (Spreyer et al. 1990). The expression pattern of apoJ/clusterin after injury to the sciatic nerve is even less well characterized. However, transcript levels for apoJ were shown to reach 2.5-fold the basal constitutive level by the end of the first week (Bonnard et al. 1997). Since apoJ was shown to induce phospholipid efflux and cholesterol release from apoE-null mice lipid-laden macrophages, with kinetics similar to apoAI (Gelissen et al. 1998), it is plausible that apoJ carries out an analogous function in the regenerating peripheral nerve.

Thus, peripheral neural tissue would have its own independent system of apolipoproteins that are responsible for the movement of lipids and for cholesterol homeostasis. ApoJ, in concert with extraneural apoAI and apoAIV, might promote the efflux of lipids from cholesterol-loaded cells. Alternatively, apoAI might deliver lipids to macrophage scavenger receptor class B-expressing cells via binding to this receptor (Xu et al. 1997). It is tempting to speculate that both apoD and apoE facilitate the cellular efflux of lipids by allowing the expansion of the cholesteryl ester-rich core of lipoprotein particles.

This process is dependent on the cholesterol-esterifying enzyme lecithin:cholesterol acyltransferase (LCAT), whose activity was found to increase in rat sciatic nerve following a crush lesion (Yao and Dyck 1981) and is perhaps modulated by apoD. The formation of an expanded cholesteryl core with a concomitant increase in the shell surface constituents (phospholipids and free cholesterol) is known to require apoE (Koo et al. 1985). Thus, the acquisition of apoE by lipoprotein particles would enable them to accommodate additional lipids and to interact with lipoprotein receptors.

4. The role of apolipoprotein E in the central nervous system in response to experimental deafferentation

4.1. Lesion-induced synaptic replacement

The dentate gyrus of the rat hippocampal formation has been used extensively as a model system to investigate the rearrangement of residual neuronal circuitry and supportive elements following deafferentation. Unilateral removal or destruction of the entorhinal cortex, which provides the major input to the granule cell dendrites of the dentate gyrus, elicits reactive growth of most residual afferents ipsilateral to the lesioned side (for reviews, see Cotman and Nieto-Sampedro 1984, Deller and Frotscher 1997). Ninety percent of the synapses in the outer two-thirds of the granule cells' dendritic tree (the hippocampal molecular layer) are lost one day after entorhinal cortex lesion (ECL). Beginning 3–6 days post-lesion (dpl), neuronal sprouts originating from septal cholinergic fibers, hippocampal CA4/hilar pyramidal (commissural-associational system) glutamatergic fibers and, to a lesser extent, contralateral entorhinal cortex fibers grow in order to replace lost synapses. These systems yield the almost complete restoration of synapses in the molecular layer within approximately two months.

A prerequisite to terminal proliferation and formation of new synapses is the clearance from the interstitium of cellular debris and of released lipids from damaged neurons (Cotman and Nieto-Sampedro 1984 and references therein; Fagan et al. 1998). Accordingly, in a manner similar to the Schwann cells and the macrophages of the peripheral nerve, both microglia and astrocytes sequentially incorporate degenerating fibers (Bechmann and Nitsch 1997). In contrast to the peripheral nerve model, hematogenous monocytes do not seem to enter areas of terminal degeneration in the hippocampus (Fagan and Gage 1994). Microglia become phagocytic as early as 1 dpl, which is followed by a delayed, but long-lasting, astrocytic hypertrophy starting at day 2 post-lesion. However, the first unequivocal signs of phagocytosis by astrocytes appear at 6 dpl (Matthews et al. 1976, Bechmann and Nitsch 1997). Both microglial cells and astrocytes were seen to contain myelinated fibers and axon terminals. Once metabolized, this cellular debris generates a large glial store of lipids, providing a convenient and readily retrievable pool of precursors for use in membrane synthesis during the formation of neuronal sprouts and the reorganization of the dendritic field of granule cell neurons. There is also evidence suggesting that phagocytic glial cells are also engaged in the remodelling of dendrites that have lost their synaptic input (Diekmann et al. 1996).

4.2. The discovery of a role for apoE in central neuron reinnervation

We have used the entorhinal cortex lesion paradigm in order to study the molecular mechanisms associated with deafferentation and reinnervation in the CNS. Differential screening of a hippocampal cDNA library from entorhinal cortex-lesioned rats led to the identification of apoE mRNA as a predominant transcript expressed in response to deafferentation (Poirier et al. 1991a). Northern blot analysis of total hippocampal RNA confirmed the increase in ApoE mRNA concentration after ECL (Poirier et al. 1991b). This increase was observed at day 2 post-lesion, the earliest time point analyzed, and was no longer apparent at 30 days. Maximum expression (7-fold above control levels) of apoE transcripts occurred around day 6 post-ECL when reactive synaptogenesis and terminal proliferation are in their early phase. ApoE mRNA was localized to hippocampal astrocytes in the deafferented zone of the molecular layer of the dentate gyrus by *in situ* hybridization in combination with immunohistochemistry for the astrocyte-specific intermediate filament protein, GFAP (Poirier et al. 1991b). ApoE immunoreactivity also peaked at 6 dpl (Poirier et al. 1991b), but the apoE protein levels increased by only 70–100% (Poirier et al. 1993b). During the first week after ECL, phosphocholine (a precursor to phosphatidylcholine or PC) levels were found to increase in homogenates of total ipsilateral hippocampus, and is thought to result from increased synthesis (Geddes et al. 1997). The time course of the elevation closely matches the time course of neuritic sprouting and perhaps reflects the PC requirements for membrane synthesis associated with axonal elongation and/or astrocytic hypertrophy (Geddes et al. 1997). A concomitant and transient accumulation of cholesterol esters and phospholipids, as measured by Sudan Black histochemistry, was seen in the outer molecular layer of the dentate gyrus and in the hilar region (Poirier et al. 1993b). By analogy to its role in the peripheral nerve following injury, apoE is presumably secreted by astrocytes in the deafferentated hippocampus in order to scavenge cholesterol and other membrane components for lipid storage and recycling to sprouting neurites via apoE receptors.

Consistent with this concept are the results we have obtained from autoradiographic analysis of the LDL binding site density in the hippocampal formation following ECL. In contrast to apoE-containing particles, LDL particles contain apoB and bind only to the LDL (apoB/E) receptor of the LDL receptor family members present in the CNS. Incubation of brain sections from ECL rats with ^{125}I-LDL or fluorescent DiI-LDL revealed an increase, compared to non-lesioned brains, in the labelling over granule cell neurons undergoing dendritic remodelling and compensatory synaptogenesis (Poirier et al. 1993b). The increase in labeling paralleled the time-course for cholinergic synaptogenesis as demonstrated by acetylcholine esterase staining profiles during the reinnervation process (Poirier et al. 1993b). The coordinated expression of apoE with the LDL receptor appears to regulate the transport of cholesterol and phospholipids during the early and intermediate phases of the reinnervation process. The relative contributions and expression patterns of other apoE receptors in this lesion model has yet to be determined.

The salvage, cellular uptake and recycling of cholesterol may explain not only the need for increased apoE expression, as observed during reactive synaptogenesis, but also the reduced levels of cholesterol synthesis as measured by 3-hydroxy-3-methylglutaryl coenzyme A (HMG-CoA) reductase activity (Poirier et al. 1993b). Following ECL, HMG-CoA reductase activity in the hippocampus was shown to progressively decline until the eighth day post-lesion, before returning to control levels by one month post-lesion (Poirier et al. 1993b). Accumulation of terminal-derived cholesterol into debris-clearing astrocytes would be consistent with the down-regulation of the HMG-CoA reductase and the LDL receptor genes which coincides with the period of maximal apoE expression. This scenario is reversed after the first week post-lesion when the process of reinnervation is well underway and during which neuronal demand for cholesterol is high. Moreover, non-esterified cholesterol levels remained relatively unchanged during the entire course of hippocampal deafferentation and reinnervation (Poirier et al. 1993b).

Taken together, these findings suggest that non-esterified cholesterol released during terminal breakdown is being recycled. Cholesterol, or its esterified form, is delivered via uncharacterized apoE-containing lipoprotein complexes to neurons undergoing reinnervation/sprouting, and taken-up through the low-density lipoprotein receptor pathway where it is presumably used as a precursor molecule for the synthesis of new synapses and terminals.

The importance of apoE in the CNS ability to compensate for deafferentation has been further investigated and highlighted by the recent work of Masliah and colleagues (1995) using apoE-deficient (knockout) homozygous mice together with control wild-type mice. These studies showed that in the apoE-deficient mice there was a delay in the patterns of reinnervation of the dentate molecular layer after perforant pathway interruption (removal of entorhinal cortex projections by aspiration). Homozygous apoE-deficient mice displayed a significant loss of synapses and disruption of the dendritic cytoskeleton and showed a poor reparative ability after the lesion.

5. Relevance of apoE mobilization of lipids to Alzheimer's disease

Although gliosis is present in the aging brain of normal and Alzheimer patients, cerebral atrophy is observed as a consequence of neuronal cell and synaptic losses, these being more considerable in the latter subjects and a characteristic of the disease. In AD, some pathways or regions in the brain are known to be more affected than others by these losses. Several of these regions, which include the entorhinal cortex and components of the brain cholinergic system, such as the medial septum and the nucleus basalis, send projections to the hippocampal formation, a brain structure known to play a major role in memory and cognitive performances. The entorhinal cortex represents the earliest site of neuropathology in both aging and AD. Denervation of the dentate gyrus, due to loss of the perforant pathway projection from the entorhinal cortex, is proposed to contribute to the pathophysiology of AD (Hyman et al. 1984, 1986). Moreover, the usual compensatory growth of fibers associated with hippocampal deafferentation appears somewhat compromised in most AD subjects. The identification of apolipoprotein E as a major responder to hippocampal formation deafferentation, as seen with the ECL model, together with the identification of the APOE ε4 allele as a major risk factor for AD, led to the hypothesis that disturbances in lipid homeostasis in the CNS could be central to Alzheimer's disease pathophysiology.

In view of apoE's role in lipid transport, it was proposed that the low levels of apoE reported in the brain of ε4 AD subjects (Bertrand et al. 1995, Pirttilä et al. 1996) may result in inefficient cholesterol and phospholipid delivery to cells in the CNS, thereby having a selective and direct impact on central cholinergic neurotransmission and synaptic integrity (Poirier 1994). Brain membrane phospholipids, particularly phosphatidylcholine (PC) and phosphatidylethanolamine (PE), are sources of choline, a rate-limiting precursor in the synthesis of acetylcholine (ACh) (Wurtman 1992). In support of this hypothesis is the finding that brain levels of choline were decreased by up to 40–50% in frontal and parietal cortices of AD patients (Nitsch et al. 1992); levels of cholesterol, which is required for the proper functioning of nicotinic receptor subtypes, were also decreased (Jones and McNamee 1988, Svennerholm and Gottfries 1994). This hypothesis is also consistent with membrane defects such as changes in membrane fluidity as observed in the hippocampus of AD subjects (Wurtman 1992).

As losses of cholinergic neurons and/or choline acetyltransferase (ChAT) activity are well known features of AD (Perry et al. 1977, Whitehouse et al. 1982), it is interesting to note that APOE ε4 allele copy number shows an inverse relationship with residual brain ChAT activity and nicotinic receptor binding sites in both the hippocampal formation and the temporal cortex of AD subjects (Poirier et al. 1995, Soininen et al. 1995, Arendt et al. 1997, Allen et al. 1997). Furthermore, the density of cholinergic neurons in the basal forebrain, which represents the primary cholinergic input to these areas, was significantly reduced in ε4 allele carriers when compared to non-ε4 AD cases or control subjects (Poirier et al. 1995, Arendt et al. 1997). In another study, the decreased metabolic activity of neurons found in the nucleus basalis of AD brains compared with that of controls was shown to be more important in apoE ε4 carriers (Salehi et al. 1998).

The fact that the ratio between the loss of cholinergic markers in the projecting areas and the loss of cholinergic neurons in the basal forebrain is lower in non-ε4 carriers suggests that the brain of these individuals can compensate to a certain extent for the loss of synapses by producing more acetylcholine and/or by sprouting. Increased cholinergic neuron vulnerability and reduced plasticity in ε4 carriers would support the hypothesis of a deficient delivery of cholesterol and phospholipids (including donor intermediates of choline) to cells in the central nervous system as a consequence of lower apoE expression in these subjects (Poirier 1994, Bertrand et al. 1995, Pirttilä et al. 1996, Beffert et al. 1999).

The animal lesion models have revealed the importance of apoE in the mobilization of lipids within nervous tissues. A link between apoE's role in lipid homeostasis in the CNS and AD could be established from the association between AD and APOE ε4 allele carriers, and the fact that brain apoE levels are lower in the latter subjects. Drug treatments aiming the upregulation of the APOE gene might thus prove to be beneficial for AD patients, especially those carrying the APOE ε4 allele. Clinical trials are in progress and preliminary data have indeed shown this approach to be feasible. As levels of cerebrospinal fluid apoE rose a concomitant improvement in cognitive functions and a slowing down of the progression of the disease was observed (Poirier, unpublished data).

Acknowledgments

Portions of the work reviewed here was conducted with the support of the Medical Research Council of Canada, the Alzheimer Society of Canada, the Fonds de la Recherche en Santé du Québec, and a sponsorship from Alcan Canada.

References

Allen, S.J., MacGowan, S.H., Tyler, S., Wilcock, G.K., Robertson, A.G.S., Holden, P.H., Smith, S.K.F. and Dawbarn, D. (1997) Reduced cholinergic function in normal and Alzheimer's disease brain is associated with apolipoprotein E4 genotype. Neurosci. Lett. 239, 33–36.

Arendt, T., Schindler, C., Brückner, M.K., Eschrich, K., Bigl, V., Zedlick, D. and Marcova, L. (1997) Plastic neuronal remodeling is impaired in patients with Alzheimer's disease carrying apolipoprotein ε4 allele. J. Neurosci. 17, 516–529.

Bechmann, I. and Nitsch, R. (1997) Astrocytes and microglial cells incorporate degenerating fibers following entorhinal lesion: a light, confocal, and electron microscopical study using a phagocytosis-dependent labeling technique. Glia 20, 145–154.

Beffert, U., Aumont, N., Dea, D., Lussier-Cacan, S., Davignon, J. and Poirier, J. (1998a) β-amyloid peptides increase the binding and internalization of apolipoprotein E to hippocampal neurons. J. Neurochem. 70, 1458–1466.

Beffert, U., Danik, M., Krzywkowski, P., Ramassamy, C., Berrada, F. and Poirier, J. (1998b) The neurobiology of apolipoproteins and their receptors in the CNS and Alzheimer's disease. Brain Res. Rev. 27, 119–142.

Beffert, U., Cohn, J.S., Petit-Turcotte, C., Tremblay, M., Aumont, N., Ramassamy, C., Davignon, J. and Poirier, J. (1999) Apolipoprotein E and beta-amyloid levels in the hippocampus and frontal cortex of Alzheimer's disease subjects are disease-related and apolipoprotein E genotype dependent. Brain Res. 843, 87–94.

Bertrand, P., Poirier, J., Oda, T., Finch, C.E. and Pasinetti, G.M. (1995) Association of apolipoprotein E genotype with brain levels of apolipoprotein E and apolipoprotein J (clusterin) in Alzheimer disease. Mol. Brain Res. 33, 174–178.

Bonnard, A.S., Chan, P. and Fontaine, M. (1997) Expression of clusterin and C4 mRNA during rat peripheral nerve regeneration. Immunopharmacology 38, 81–86.

Borghini, I., Barja, F., Pometta, D. and James, R.W. (1995) Characterization of subpopulations of lipoprotein particles isolated from human cerebrospinal fluid. Biochim. Biophys. Acta 1255, 192–200.

Boyles, J.K., Pitas, R.E., Wilson, E., Mahley, R.W. and Taylor, J.M. (1985) Apolipoprotein E associated with astrocytic glia of the central nervous system and with nonmyelinating glia of the peripheral nervous system. J. Clin. Invest. 76, 1501–1513.

Boyles, J.K., Zoellner, C.D., Anderson, L.J., Kosik, L.M., Pitas, R.E., Weisgraber, K.H., Hui, D.Y., Mahley, R.W., Gebicke-Haerter, P.J., Ignatius, M.J. and Shooter, E.M. (1989) A role for apolipoprotein E, apolipoprotein A-I, and low density lipoprotein receptors in cholesterol transport during regeneration and remyelination of the rat sciatic nerve. J. Clin. Invest. 83, 1015–1031.

Boyles, J.K., Notterpek, L.M. and Anderson, L.J. (1990) Accumulation of apolipoproteins in the regenerating and remyelinating mammalian peripheral nerve. J. Biol. Chem. 265, 17805–17815.

Corder, E.H., Saunders, A.M., Strittmatter, W.J., Schmechel, D.E., Gaskell, P.C., Small, G.W., Roses, A.D., Haines, J.L. and Pericak-Vance, M.A. (1993) Gene dose of apolipoprotein E type 4 allele and the risk of Alzheimer's disease in late onset families. Science 261, 921–923.

Cotman, C.W. and Nieto-Sampedro, M. (1984) Cell biology of synaptic plasticity. Science 225, 1287–1294.

Das, H.K., McPherson, J., Bruns, G.A.P., Karathanasis, S.K. and Breslow, J.L. (1985) Isolation, characterization, and mapping to chromosome 19 of the human apolipoprotein E gene. J. Biol. Chem. 260, 6240–6247.

Davignon, J., Gregg, R.E. and Sing, C.F. (1988) Apolipoprotein E polymorphism and atherosclerosis. Arteriosclerosis 8, 1–21.

Deller, T. and Frotscher, M. (1997) Lesion-induced plasticity of central neurons: sprouting of single fibres in the rat hippocampus after unilateral entorhinal cortex lesion. Prog. Neurobiol. 53, 687–727.

Diedrich, J.F., Minnigan, H., Carp, R.I., Whitaker, J.N., Race, R., Frey II, W. and Haase, A.T. (1991) Neuropathological changes in Scrapie and Alzheimer's disease are associated with increased expression of apolipoprotein E and cathepsin D in astrocytes. J. Virol. 65, 4759–4768.

Diekmann, S., Ohm, T.G. and Nitsch, R. (1996) Long-lasting transneuronal changes in rat dentate granule cell dendrites after entorhinal cortex lesion. A combined intracellular injection and electron microscopy study. Brain Pathol. 6, 205–215.

Elshourbagy, N.A., Liao, W.S., Mahley, R.W. and Taylor, J.M. (1985) Apolipoprotein E mRNA is abundant in the brain and adrenals as well as in the liver, and is present in other peripheral tissues of rats and marmosets. Proc. Natl. Acad. Sci. U.S.A. 82, 203–207.

Fagan, A.M. and Gage, F.H. (1994) Mechanisms of sprouting in the adult central nervous system: cellular responses in areas of terminal degeneration and reinnervation in the rat hippocampus. Neurosci. 58, 705–725.

Fagan, A.M., Murphy, B.A., Patel, S.N., Kilbridge, J.F., Mobley, W.C., Bu, G. and Holtzman, D.M. (1998) Evidence for normal aging of the septo-hippocampal cholinergic system in apoE(−/−) mice but impaired clearance of axonal degeneration products following injury. Exp. Neurol. 151, 314–325.

Farrer, L.A., Cupples, L.A., Haines, J.L., Hyman, B., Kukull, W.A., Mayeux, R., Myers, R.H., Pericak-Vance, M.A., Risch, N. and van Duijn, C.M. (1997) Effects of age, sex, and ethnicity on the association between apolipoprotein E genotype and Alzheimer Disease: A meta-analysis. J. Am. Med. Assoc. 278, 1349–1356.

Geddes, J.W., Panchalingam, K., Keller, J.N. and Pettigrew, J.W. (1997) Elevated phosphocholine and phosphatidylcholine following rat entorhinal cortex lesions. Neurobiol. Aging 18, 305–308.

Gelissen, I.C., Hochgrebe, T., Wilson, M.R., Easterbrooksmith, S.B., Jessup, W., Dean, R.T. and Brown, A.J. (1998) Apolipoprotein J (clusterin) induces cholesterol export from macrophage-foam cells: a potential anti-atherogenic function. Biochem. J. 331, 231–237.

Gliemann, J. (1998) Receptors of the low density lipoprotein (LDL) receptor family in man. Multiple functions of the large family members via interaction with complex ligands. Biol. Chem. 379, 951–964.

Goodrum, J.F. (1990) Cholesterol synthesis is down-regulated during regeneration of peripheral nerve. J. Neurochem. 54, 1709–1715.

Goodrum, J.F. (1991) Cholesterol from degenerating nerve myelin becomes associated with lipoproteins containing apolipoprotein E. J. Neurochem. 56, 2082–2086.

Goodrum, J.F. and Bouldin, T.W. (1996) The cell biology of myelin degeneration and regeneration in the peripheral nerve. J. Neuropathol. Exp. Neurol. 55, 943–953.

Goodrum, J.F., Weaver, J.E., Goines, N.D. and Bouldin, T.W. (1995) Fatty acids from degenerating myelin lipids are conserved and reutilized for myelin synthesis during regeneration in peripheral nerve. J. Neurochem. 65, 1752–1759.

Han, S.H., Einstein, G., Weisgraber, K.H., Strittmatter, W.J., Saunders, A.M., Pericak-Vance, M.A., Roses, A.D. and Schmechel, D.E. (1994) Apolipoprotein E is localized to the cytoplasm of human cortical neurons: a light and electron microscopic study. J. Neuropathol. Exp. Neurol. 53, 535–544.

Higgins, G.A., Large, C.H., Rupniak, H.T. and Barnes, J.C. (1997) Apolipoprotein E and Alzheimer's disease: a review of recent studies. Pharmacol. Biochem. Behav. 56, 675–685.

Hyman, B.T., Van Horsen, G.W., Damasio, A.R. and Barnes, C.L. (1984) Alzheimer's disease: cell-specific pathology isolates the hippocampal formation. Science 225, 1168–1170.

Hyman, B.T., Van Horsen, G.W., Kromer, L.J. and Damasio, A.R. (1986) Perforant pathway changes and the memory impairment of Alzheimer's disease. Ann. Neurol. 20, 472–481.

Ignatius, M.J., Gebicke-Haerter, P.J., Skene, J.H.P., Schilling, J.W., Weisgraber, K.H., Mahley, R.W. and Shooter, E.M. (1986) Expression of apolipoprotein E during nerve degeneration and regeneration. Proc. Natl. Acad. Sci. U.S.A. 83, 1125–1129.

Jones, O.T. and McNamee, M.G. (1988) Annular and nonannular binding sites for cholesterol associated with the nicotinic acetylcholine receptor. Biochemistry 27, 2364–2374.

Koo, C., Innerarity, T.L. and Mahley, R.W. (1985) Obligatory role of cholesterol and apolipoprotein E in the formation of large cholesterol-enriched and receptor-active high density lipoproteins. J. Biol. Chem. 260, 11934–11943.

LaDu, M.J., Gilligan, S.M., Lukens, J.R., Cabana, V.G., Reardon, C.A., Van Eldik, L.J. and Holtzman, D.M. (1998) Nascent astrocyte particles differ from lipoproteins in CSF. J. Neurochem. 70, 2070–2081.

Mahley, R.W. (1988) Apolipoprotein E: cholesterol transport protein with expanding role in cell biology. Science 240:622–630.

Masliah, E., Mallory, M., Ge, N., Alford, M., Veinbergs, I. and Roses, A.D. (1995) Neurodegeneration in the central nervous system of apoE-deficient mice. Exp. Neurol. 136, 107–122.

Matthews, D.A., Cotman, C.W. and Lynch, G. (1976) An electron microscopic study of lesion-induced synaptogenesis in the dentate gyrus of the adult rat. Brain Res. 115, 1–41.

Nakai, M., Kawamata, T., Tanigushi, T., Maeda, K. and Tanaka, C. (1996) Expression of apolipoprotein E mRNA in rat microglia. Neurosci. Lett. 211, 41–44.

Nalbantoglu, J., Gilfix, B.M., Bertrand, P., Robitaille, Y., Gauthier, S., Rosenblatt, S. and Poirier, J. (1994) Predictive value of apolipoprotein E genotyping in Alzheimer's disease. Ann. Neurol. 36, 889–895.

Nitsch, R.M., Blusztajn, J.K., Pittas, A.G., Slack, B.E., Growdon, J.H. and Wurtman, R.J. (1992) Evidence for a membrane defect in Alzheimer disease brain. Proc. Natl. Acad. Sci. U.S.A. 89, 1671–1675.

Osman, I., Gaillard, O., Meillet, D., Bordas-Fonfrede, M., Gervais, A., Schuller, E., Delattre, J. and Legrand, A. (1995) A sensitive time-resolved immunofluorometric assay for the measurement of apolipoprotein B in cerebrospinal fluid. Application to multiple sclerosis and other neurological diseases. Eur. J. Clin. Chem. Clin. Biochem. 33, 53–58.

Perry, E.K., Gibson, P.H., Blessed, G., Perry, R.H. and Tomlinson, B.E. (1977) Neurotransmitter enzyme abnormalities in senile dementia. Choline acetyltransferase and glutamic acid decarboxylase activities in necropsy brain tissue. J. Neurol. Sci. 34, 247–265.

Pirttilä, T., Soininen, H., Heinonen, O., Lehtimäki, T., Bogdanovic, N., Paljarvi, L., Kosunen, O., Winblad, B., Riekkinen, P., Wisniewski, H.M. and Mehta, P.D. (1996) Apolipoprotein E (apoE) levels in brains from Alzheimer disease patients and controls. Brain Res. 722, 71–77.

Pitas, R.E., Boyles, J.K., Lee, S.H., Foss, D. and Mahley, R.W. (1987) Astrocytes synthesize apolipoprotein E and metabolize apolipoprotein E-containing lipoproteins. Biochim. Biophys. Acta 917, 148–161.

Poirier, J. (1994) Apolipoprotein E in animal models of CNS injury and in Alzheimer's disease. Trends Neurosci. 17, 525–530.

Poirier, J., Hess, M., May, P.C. and Finch, C.E. (1991a) Cloning of hippocampal poly(A) RNA sequences that increase after entorhinal cortex lesion in adult rat. Mol. Brain Res. 9, 191–195.

Poirier, J., Hess, M., May, P.C. and Finch, C.E. (1991b) Astrocytic apolipoprotein E mRNA and GFAP mRNA in hippocampus after entorhinal cortex lesioning. Mol. Brain Res. 11, 97–106.

Poirier, J., Davignon, J., Bouthillier, D., Kogan, S., Bertrand, P. and Gauthier, S. (1993a) Apolipoprotein E polymorphism and Alzheimer's disease. Lancet 342, 697–699.

Poirier, J., Baccichet, A., Dea, D. and Gauthier, S. (1993b) Cholesterol synthesis and lipoprotein reuptake during synaptic remodelling in hippocampus in adult rats. Neurosci. 55, 81–90.

Poirier, J., Delisle, M.C., Quirion, R., Aubert, I., Farlow, M., Lahiri, D., Hui, S., Bertrand, P., Nalbantoglu, J., Gilfix, B.M. and Gauthier, S. (1995) Apolipoprotein E4 allele as a predictor of cholinergic deficits and treatment outcome in Alzheimer disease. Proc. Natl. Acad. Sci. U.S.A. 92, 12260–12264.

Rall Jr, S.C., Weisgraber, K.H. and Mahley, R.W. (1982) Human apolipoprotein E. The complete amino acid sequence. J. Biol. Chem. 257, 4171–4178.

Rawlins, F.A., Hedley-White, E.T., Villegas, G. and Uzman, B.G. (1970) Reutilization of cholesterol-1,2-^{3}H in the regeneration of peripheral nerve. Lab. Invest. 22, 237–240.

Rubinsztein, D.C. (1995) Apolipoprotein E: a review of its roles in lipoprotein metabolism, neuronal growth and repair and as a risk factor for Alzheimer's disease. Psychol. Med. 25, 223–229.

Salehi, A., Dubelaar, E.J.G., Mulder, M. and Swaab, D.F. (1998) Aggravated decrease in the activity of nucleus basalis neurons in Alzheimer's disease is apolipoprotein E-type dependent. Proc. Natl. Acad. Sci. U.S.A. 95, 11445–11449.

Saunders, A.M., Strittmatter, W.J., Schmechel, D.E., St. George-Hyslop, P.H., Pericak-Vance, M.A., Joo, S.H., Rosi, B.L., Gusella, J.F., Crapper-MacLachlan, D.R., Alberts, M.J., Hulette, C., Crain, B., Goldgaber, D. and Roses, A.D. (1993) Association of apolipoprotein E allele epsilon 4 with late-onset familial and sporadic Alzheimer's disease. Neurology 43, 1467–1472.

Schmechel, D.E., Tiller, M.O., Tong, P., McSwain, M., Han, S.H., Ange, R., Burkhart, D.S. and Izard, M.K. (1996) Pattern of apolipoprotein E immunoreactivity during brain aging. In: A.D. Roses, K.H. Weisgraber and Y. Christen (Eds.), Apolipoprotein E and Alzheimer's Disease, Springer, Berlin, pp. 29–48.

Snipes, G.J., McGuire, C.B., Norden, J.J. and Freeman, J.A. (1986) Nerve injury stimulates the secretion of apolipoprotein E by non-neuronal cells. Proc. Natl. Acad. Sci. U.S.A. 83, 1130–1134.

Soininen, H., Kosunen, O., Helisalmi, S., Mannermaa, A., Paljärvi, L., Talasniemi, S., Ryynänen, M. and Riekkinen Sr, P. (1995) A severe loss of choline acetyltransferase in the frontal cortex of Alzheimer patients carrying apolipoprotein ε4 allele. Neurosci. Lett. 187, 79–82.

Spreyer, P., Schaal, H., Kuhn, G., Rothe, T., Unterbeck, A., Olek, K. and Müller, H.W. (1990) Regeneration-associated high level expression of apolipoprotein D mRNA in endoneurial fibroblasts of peripheral nerve. EMBO J. 9, 2479–2484.

Stone, D.J., Rozovsky, I., Morgan, T.E., Anderson, C.P., Hajian, H. and Finch, C.E. (1997) Astrocytes and microglia respond to estrogen with increased apoE mRNA in vivo and in vitro. Exp. Neurol. 143, 313–318.

Strittmatter, W.J., Saunders, A.M., Schmechel, D.E., Pericak-Vance, M.A., Enghild, J., Salvesen, G.S. and Roses, A.D. (1993) Apolipoprotein E: high-avidity binding to beta-amyloid and increased frequency of type 4 allele in late-onset familial Alzheimer disease. Proc. Natl. Acad. Sci. U.S.A. 90, 1977–1981.

Svennerholm, L. and Gottfries, C.G. (1994) Membrane lipids, selectively diminished in Alzheimer brains, suggest synapse loss as a primary event in early-onset form (type I) and demyelination in late-onset form (type II). J. Neurochem. 62, 1039–1047.

Utermann, G., Pruin, N. and Steinmetz, A. (1979) Apolipoprotein E polymorphism in health and disease. Clin. Genet. 15, 37–62.

Weiler-Guttler, H., Sommerfeldt, M., Papandrikopoulou, A., Mischek, U., Bonitz, D., Frey, A., Grupe, M., Scheerer, J. and Gassen, H.G. (1990) Synthesis of apolipoprotein A-1 in pig brain microvascular endothelial cells. J. Neurochem. 54, 444–450.

White, F.V., Toews, A.D., Goodrum, J.F., Novicki, D.L., Bouldin, T.W. and Morell, P. (1989) Lipid metabolism during early stages of Wallerian degeneration in the rat sciatic nerve. J. Neurochem. 52, 1085–1092.

Whitehouse, P.J., Price, D.L., Struble, R.G., Clark, A.W., Coyle, J.T. and Delon, M.R. (1982) Alzheimer's disease and senile dementia: loss of neurons in the basal forebrain. Science 215, 1237–1239.

Wood, J.G. and Dawson, R.M.C. (1974) Lipid and protein changes in sciatic nerve during Wallerian degeneration. J. Neurochem. 22, 631–635.

Wurtman, R.J. (1992) Choline metabolism as a basis for the selective vulnerability of cholinergic neurons. Trends Neurosci. 15, 117–122.

Xu, S., Laccotripe, M., Huang, X., Rigotti, A., Zannis, V.I. and Krieger, M. (1997) Apolipoproteins of HDL can directly mediate binding to the scavenger receptor SR-BI, an HDL receptor that mediates selective lipid uptake. J. Lipid Res. 38, 1289–298.

Yao, J.K. and Dyck, P.J. (1981) Cholesterol esterifying enzyme in normal and degenerating peripheral nerve. J. Neurochem. 37, 156–163.

Yao, J.K., Natarajan, V. and Dyck, P.J. (1980) The sequential alterations of endoneurial cholesterol and fatty acid in Wallerian degeneration and regeneration. J. Neurochem. 35, 933–940.

Zannis, V.I. and Breslow, J.L. (1981) Human VLDL-apoE isoprotein polymorphism is explained by genetic variation and post-translational modifications. Biochemistry 20, 1033–1041.

E.R. Skinner (Ed.), *Brain Lipids and Disorders in Biological Psychiatry*
Published by Elsevier Science B.V.

CHAPTER 5

Omega-3 fats in depressive disorders and violence: the context of evolution and cardiovascular health

Joseph R. Hibbeln and Kevin K. Makino

National Institute on Alcohol Abuse and Alcoholism, NIH, Bethesda, MD, USA

1. Introduction

The study of omega-3 essential fatty acids in affective disorders remains a field in its infancy. A major impetus to this field of study has been the observation that docosahexaenoic acid (DHA) is selectively concentrated in synaptic neuronal membranes. Thus in our original hypothesis paper we attempted to address the following broad question: what, if any, are the psychiatric consequences of insufficient levels of DHA in the brain? (Hibbeln and Salem 1995). Specifically, is an insufficiency of DHA or omega-3 fatty acids related to an increased predisposition to suffering depressive symptoms? We will also examine if an insufficiency of omega-3 fatty acids is associated with other affective disorders such as postpartum depression, bipolar affective disorder and impulsivity in homicide and suicide. The major focus of this chapter will be the broader evolutionary and epidemiological context of the following question: does it make reasonable sense that an insufficiency of omega-3 fatty acids, including both DHA and eicosapentaenoic acid (EPA), would be associated with depression? A critical focus of this question will be the hypothesis that an insufficiency of omega-3 fatty acids may be responsible for the observations that depression and hostility are linked to a greater risk of cardiovascular disease. There does not appear to be any evolutionary advantage to either suffering major depression or early cardiovascular death. While it would be intellectually satisfying to determine what neurochemical or neurophysiological mechanism is responsible for this link, the question of determining a specific mechanism is premature. Not only are pathophysiological mechanisms underlying depression not well understood, but the biological and biophysical necessity for DHA in the central nervous system remains an unresolved question. None the less, preliminary mechanistic hypotheses of serotonergic function, Hypothalamic–Pituitary–Adrenal axis function and immune neuroendocrine mechanisms will be briefly explored in this chapter. Finally, we will contrast the data regarding omega-3 fatty acids with the hypothesis that lowering plasma cholesterol concentrations (or endogenously low cholesterol concentrations) increase affective symptoms including depression and impulsiveness, which could lead to homicide or suicide.

Several factors have made this field of study attractive to patients and practicing clinicians, including the non-toxic and nutritionally based approach to treatment along with lack of known adverse interactions with other psychotropic medications. However, an important consideration is that physicians are already using these compounds in their practices with good results. There is a great diversity in the etiology and prognosis of the various depressive disorders, some of which may respond to α-linolenic acid (LNA), EPA and/or to DHA while others may not. Because this field is only beginning to emerge, this chapter will review data regarding the relationships of EPA and DHA to psychiatric disorders that are available from meeting abstracts, private communications and unpublished data in addition to published data. We hope to provide a broad base of information for consideration. Perhaps it is useful to contrast the infancy of this field to the role of omega-3 fats in cardiovascular disorders. It was more than 20 years ago the study of EPA and DHA in cardiovascular disorders was stimulated by epidemiological findings of lower rates of cardiovascular disease among Inuit Eskimos compared to Danish populations (Dyerberg and Bang 1979, Menotti et al. 1989, Kromhout 1989). However, it has only been within the last year that the American Heart Association has included the consumption of omega-3 fatty acids in its dietary recommendations (Krauss et al. 2000). Cross-national comparisons of omega-3 fatty acids in psychiatric disorders are only beginning to emerge (Hibbeln 1998). With luck, the field of omega-3 fatty acids in psychiatry will follow the successes of omega-3 fatty acids in cardiovascular disorders.

1.1. Major depression in the context of evolution

Fundamental to the understanding of any biological process, including mental illnesses, is an understanding of its evolutionary context. There are few readily apparent evolutionary advantages for the development of a high prevalence of major depressive illnesses among *Homo sapiens*. Major depression frequently destroys people's lives and has a chronic recurring course. Major depression is defined by DSM-IV (American Psychiatric Association 1994) by the loss of the ability to function in family or job life for at least two weeks, due to disturbances in mood, sleep, concentration, self-esteem, appetite, physical energy and sexual energy or function. This disease not only directly reduces the likelihood of procreation, but also causes significant disruption in social and family interactions. Compared to current rates, prevalence rates of major depression may have been nearly 100-fold lower prior to 1910 (Klerman and Weissman 1989). These increased rates of prevalence of affective disorders that have been identified among birth cohorts during the last century in the United States have been well documented (Klerman and Weissman 1989, Robins and Regier 1991, Wickramaratne et al. 1989). These increased prevalence rates of depressive disorders during the last century have also been described in nine other industrialized nations (Cross-National Collaborative Group 1992, Klerman and Weissman 1989). The argument that this cohort effect arose from changes in gene frequency over these few decades has been questioned, and other potential explanations that have been offered include nutritional, environmental and social causes (Bebbington and Ramana 1995).

Several investigators (Broadhurst et al. 1998, Eaton and Konner 1985, Eaton et al. 1998) have postulated that our genetic patterns evolved in the context of a nutritional milieu that allowed for optimal brain development and psychiatric functioning. Unfortunately, with regards to essential fatty acid intake, the modern diets of post-industrialized societies appear to be discordant with our genetic pattern and may contribute to the increased prevalence rates of major depression (Hibbeln and Salem 1995). *Homo sapiens* are thought to have evolved consuming diets rich in directly available long-chain omega-3 fatty acids, which may have been permissive for the development of proportionally larger brains (Broadhurst et al. 1998, Eaton et al. 1998). This proposition is supported by the observation that DHA is selectively concentrated at synaptic neuronal membranes and is critical for optimal neuronal function (Salem et al. 1986). DHA also stimulates neurite outgrowth in PC12 cells (Ikemoto et al. 1997) and is selectively accumulated by synaptic growth cones during neuronal development (Austead et al. 2000, Martin 1998). Walter and colleagues (Walter et al. 2000) describe that *Homo sapiens* emerged from Africa along a coastal route and include as evidence the finding of stone tools amidst oyster shells and crab claws in a 200 000 year-old coral reef. It is quite reasonable to conclude that these early coastal *Homo sapiens* frequently ate seafood and routinely delivered preformed DHA to the brain during development and during adulthood. However, during the last century diets of industrialized societies have significantly diverged from these traditional diets, reducing the absolute amounts of EPA and DHA intake while increasing the relative intake of omega-6 fatty acids, which compete with omega-3 functions (Eaton et al. 1998). The ratios of omega-6 to omega-3 fatty acids are estimated to be between 0.4 and 2.8 in Paleolithic and evolutionary diets including models in which either plants and animal butchering were considered as sole sources of food. These models did not include seafood consumption, which would bring the ratios even lower. In contrast, current average ratios are estimated to be approximately 17 in typical western industrialized societies (Eaton et al. 1998) and we have observed schizophrenics with arachidonic acid (AA) to EPA (AA/EPA) ratios in red blood cells of more than 70:1 (unpublished data). These changes in the dietary intake are based not only on decreased seafood consumption but also on the increased production of seed oils, in particular soybean and corn oils. These seed oils have much higher ratios of linoleic to alfa-linolenic acids compared to leafy plants, wild game and seafood (Sinclair et al. 1987, Naughton et al. 1986). Since livestock are also dependent upon plants for their sources of essential fatty acids, these changes have pervaded the entire food chain (Simopoulos 1998). Clearly there have been numerous dramatic changes in human civilization in the last century that may have contributed to the increased prevalence of major depression. However, since changes in the dietary intake of essential fatty acids appear to be able to directly influence central nervous system function, these nutritional factors should be carefully considered (Hibbeln and Salem 1995). Quantitative assessment of this hypothesis is unfortunately difficult due to the uneven quality and paucity of historical dietary data.

1.2. Cross-national epidemiology and major depression

One of the most compelling aspects of this field comes with an appreciation of the magnitude of the burden of major depressive illnesses. The World Health Organization has identified major depression as the world's single greatest cause of morbidity, defined as years of life lost to a disability (Virkkunen et al. 1987). Cross-national comparisons of the dietary intake of seafood and the prevalence rates of the major psychiatric illnesses can be used to estimate the magnitude of the effects of low omega-3 status upon diverse populations. It is remarkable that the prevalence rates across countries for three affective disorders, major depression, bipolar depression and postpartum depression, are each robustly related to amounts of seafood consumed in each country across 50–60-fold differences in prevalence (Hibbeln 1998, and unpublished data).

The analyses of cross-national epidemiological data, collected using high-quality modern diagnostic and epidemiological sampling methods, does provide one method of testing the hypothesis that a lower omega-3 fatty acid status is related to higher prevalence rates of affective disorders, psychotic disorders or aggressive behaviors. Economic data describing seafood consumption have been useful in these cross-national studies. While economic data on the production and consumption of seafood cannot accurately be used to quantify dietary intake for an individual, these data can be used to describe trends for entire countries, and thus provide a basis for comparing consumption cross-nationally (World Health Organization 1996). The financial incentive to produce accurate data also adds some confidence to the accuracy of consumption estimates derived from economic data. When compared cross-nationally, greater amounts of seafood consumption were robustly correlated ($r = -0.85$, $p < 0.0005$) with lower lifetime prevalence rates of major depression (Hibbeln 1998). The prevalence rates of major depression varied nearly 50-fold across countries as assessed from raw data from each country merged with the Epidemiological Catchment Area study, a gold standard of studies in psychiatric epidemiology. This approach was specifically designed for cross-national analyses and weighted for age, sex and other demographic differences (Weissman et al. 1996). These data appear to be robust after consideration of potential differences in cultural biases in diagnosis because a structured interview with standardized diagnostic criteria was validated in each country prior to the community sampling. Comparative relationships to prevalence rates of postpartum depression, bipolar affective disorder, homicide mortality, suicide mortality and schizophrenia will be presented later in this article with a discussion of each of these issues. While cross-national studies do not provide direct evidence of causal relationships, they do offer a perspective on the potential magnitude of these relationships.

The results of cross-national analysis comparing prevalence rate of major depression are consistent with results of epidemiological studies conducted within single countries. Tanskanen et al. (2001b) studied a population of 1767 subjects within Northern Finland and reported that subjects who consumed fish twice a week or more were at lower risk of reporting depressive symptoms (odds ratio 0.63) and suicidal thinking (odds ratio 0.57), compared to infrequent fish consumers. Consistent with this report, Silvers and Scott (2000) also found that greater fish consumption predicted improved mental health status by self-report among 4644 subjects in New Zealand. Unfortunately, direct data on depression was not available. An assessment of essential fatty acid status using direct tissue sampling has also yielded similar results in a community sample of 200 elderly subjects selected from a sample population of 4500 subjects representing 80% of the elderly people in two counties in Iowa (Hibbeln et al., unpublished data). Comparisons were made between the 50 most depressed men; the 50 most depressed women; 50 control men and 50 control women. Lower plasma DHA concentrations were associated with 6.5-fold greater risk of depression among women. Low plasma concentrations of plasma DHA alone also significantly predicted more severe sleep complaints and reports of anxiety among all groups (Hibbeln et al., unpublished data). DHA was the only plasma fatty acid measure or plasma cholesterol measure that was related to psychiatric symptoms. These cross-national and within-country observational studies are consistent with the hypothesis that low omega-3 status is associated with depressive symptoms.

1.3. Major depression and tissue compositional studies

Lipid compositional data from depressed patients was sparse prior to the publication of our original hypothesis. Sengupta et al. (1981) documented decreases in total serum phosphatidylserine in unipolar depressives. Abnormalities in biophysical measures of membrane order have been reported in erythrocytes, lymphocytes and platelets of patients with affective illnesses (Pettegrew et al. 1982, Piletz et al. 1991) and may have resulted from alterations in lipid composition. Two early reports (Ellis and Sanders 1977, Fehily et al. 1981) found that the levels of 22:6 n-3 in plasma phosphatidylcholine increased with severity of depressive symptoms, but these findings are difficult to interpret due to the diagnostic heterogeneity of the patients, lack of dietary assessment and reported decreases in total serum phosphatidylcholine in unipolar depressives (Sengupta et al. 1981). In addition, one third of these patients were on psychotropic medications which may cause phosholipidoses (Kodavanti and Mehendale 1990). Kodavanti and Mehendale (1990) comprehensively review the effects of cationic ampathiphillic drugs, including many psychotropic medications, on membrane lipid composition.

Since the publication of our initial hypothesis (Hibbeln and Salem 1995), a series of clinical studies have reported that depressed patients have low tissue concentrations of EPA and/or DHA, and several supplementation trials have reported improvements in depressive symptoms in the affective spectrum. Seven studies have reported that lower concentrations of n-3 fatty acids in plasma or red blood cells predicted depressive symptoms (Adams et al. 1996, Edwards et al. 1998a,b, Maes et al. 1996, 1999, Peet et al. 1998, Hibbeln et al., unpublished data). Adams and colleagues were the first to report that lower measures of DHA in the phospholipids of red blood cells ($r = 0.80$, $p < 0.01$) and a greater AA/EPA ratio ($r = 0.73$, $p < 0.01$) predicted more severe depressive symptoms (Adams et al. 1996). Several investigators have also reported elevated AA/EPA ratios among depressed subjects. This suggests that depressed patients are more likely to generate AA-derived eicosanoids, thereby increasing cardiovascular risk. For example, Maes and colleagues found lower concentrations of LNA and EPA and a higher AA/EPA ratio comparing 36 patients with major depression to 14 subjects with minor depression and to 24 healthy controls (Maes et al. 1996). One well-done study carefully controlled common confounding factors that would alter omega-3 status among depressed subjects by controlling for both alcohol consumption and cigarette smoking while also assessing dietary intake (Edwards et al. 1998b). DHA concentrations in RBCs emerged as the most significant predictor of Beck Depression Inventory Scores in a multiple regression model (beta coefficient of -0.92). These investigators (Edwards et al. 1998b) also reported lower concentrations of EPA, DHA, and total omega-3 fatty acids in red blood cells in 10 depressed patients compared to 14 healthy controls. Within those same depressed patients, severity of depression was best predicted by dietary LNA ($r = -0.83$, $p < 0.003$) and red blood cell LNA ($r = -0.81$, $p < 0.008$). Interestingly, no differences in dietary intake of omega-3 fats were found between patients and controls during the time period studied. In a carefully controlled examination of 15 unmedicated patients with major depression, Peet and colleagues found considerably reduced ($p < 0.009$) DHA concentrations in red blood cell membranes relative to 15 matched controls (Peet et al. 1998). This finding is consistent with the proposition that decreased concentrations of DHA in red blood cell membranes may reflect decreased DHA concentrations in depressed patients' neuronal membranes, altering membrane biophysical properties and affecting membrane bound receptors and enzymes found to be dysfunctional in depression (Hibbeln and Salem 1995).

Because the tissue-compositional studies have been observational, it has been difficult to absolutely determine which omega-3 fatty acids, or combination of fatty acids, have the most important mechanistic role. It is important to note that the robust correlational relationships described in these tissue-compositional studies are remarkably consistent with the robust cross-national relationships that have been described between lower seafood consumption and higher prevalence rates of major depression ($r = -0.84$, $p < 0.005$) (Hibbeln 1998). These data do offer strong support that DHA and omega-3 status is compromised among depressed subjects, but these data should not be interpreted as evidence of a metabolic defect in omega-3 essential fatty acid metabolism among depressed subjects. Since most of these studies did not assess dietary intake, the possibility that being depressed has altered dietary preferences and reduced omega-3 intake, must also be considered. Although the primary source of omega-3 insufficiency among depressed patients is likely to be poor dietary intake, other possibilities such as metabolic differences cannot be dismissed. A possible contributing factor could be increased degradation of these polyunsaturated fatty acids through oxidation caused by smoking and excessive alcohol consumption (Morrow et al. 1995, Shahar et al. 1999). Prolonged periods of psychological stress may also degrade tissue polyunsaturated fatty acid concentrations (see the review by Hibbeln and Salem 1995). Regardless of the etiology of the insufficiency, increased dietary intake can readily improve omega-3 status.

1.4. Historical observations on the treatment of depression with lipids: Considerations regarding Christianity

In western cultures, the fish is a core and ancient symbol of Christianity, which is older than the crucifix. The identity is so close that the Greek word for fish, IXΘΥΣ, is an anagram for the word Christ. The core sermon of Christianity, the sermon on the mount, describes a philosophy of non-impulsive and calm behavior. Christians are literally instructed to turn the other cheek when assaulted (Matt 5:39), which would certainly reduce homicides. Among the multiple references to fish in the bible, fish was reported to be the first meal that Jesus ate after his resurrection (Luke 24:42) before opening up the minds of disciples to understand the scriptures. Two miracles describe the multiplication of fishes for multitudes of 4000 (Mark 8:8, Matt 15:36–37) and 5000 (Luke 9:16–17, Mark 6:41, Matt 14:19–20, John 6:11), while a third describes Simon casting and filling the nets to feed the disciples after resurrection (Luke 9:16–17, John 21:6–11). This symbolism is more than a dietary recommendation to consume more fish; it is nearly a dictum to encourage production and consumption. From

1542 until the 1960s the Roman Catholic Church encouraged the consumption of fish and seafood on at least one day of the week, Friday or Wednesday or both, on all Christian feast days and throughout nearly all of the 40 days of lent. If fish were consumed on all of these days, then certainly an average of more than 3 fish meals per week would be consumed, a reasonable modern dietary recommendation. Is it possible that one reason the symbolic association of fish and peaceful behaviors has persisted for more than 2000 years in western society is that consumption of the omega-3 fatty acids in fish have the biological psychotropic effects of reducing impulsive and violent behaviors?

Some historical evidence suggests that essential fatty acids may have some treatment efficacy in depression. In *The Anatomy of Melancholy*, published in 1652, Burton recommended the use of borage oil, high in n-6 polyunsaturates, as well as a low-fat diet including fish (Burton 1988). For severe cases, Burton recommended a 2-week diet of brains, an excellent source of 22:5 n-3, 22:6 n-3 and 20:4 n-6. In a series of fascinating case reports, D.O. Rudin was perhaps the first modern physician to note a beneficial psychotropic effect of linseed oil (which is high in 18:3 n-3) among a heterogeneous mixture of patients including severely ill schizophrenics (Rudin 1981). Another curious finding has been that treatment with oral phosphatidylserine, prepared from bovine cortex (BC-PS), markedly improved depressive symptomatology in 11 elderly women (Maggioni et al. 1990). Bovine gray matter phosphatidylserine contains 29% of its fatty acids as 22:6 n-3 (Salem et al. 1980). Treatment with 300 mg/day of BC-PS markedly reduced withdrawal and apathy scores compared to a corn oil placebo, in a multi-center study of 494 elderly patients (Cenacchi et al. 1993). Withdrawal and apathy scores are often difficult to distinguish from mild depressive symptoms. Stockert and colleagues suggested BC-PS may regulate the serotonin reuptake site, a site of action of fluoxetine (Prozac®) (Stockert et al. 1989). They noted that BC-PS reduced [3H] imipramine binding sites in rat brain by 23% alone and by 47% when combined with amitryptaline. Block and Edwards described regulation of the serotonin reuptake protein by altering the biophysical properties with 18:1 n-9 and cholesterol (Block and Edwards 1987). In a difficult to interpret series of human studies, mixtures of hypothalamic phospholipids potentiated antidepressant actions (Casacchia et al. 1982, Gianelli et al. 1989, Roccatagliata et al. 1978). These lipids were also reported to increased cerebrospinal fluid 5-hydroxyindoleacetic acid (5-HIAA) and homovanillic (HVA) acid levels as well as somatotrophin release, measures often associated with an improvement in depressive symptoms (Drago et al. 1985). Unfortunately these human studies have not been adequately replicated. Similar phospholipid mixtures reduced immobility of rats in the Porsolt swim test (Mills and Ward 1989). Borage oil and 20:5 n-3 reversed psychosocial stress-induced hypertension in humans and rats (Mills et al. 1989, 1994, Mills and Ward 1989). These fragmentary but suggestive reports constitute part of the historical background of this field.

1.5. Current treatment of major depressive symptoms and bipolar affective disorder

One of the most important contributions to the field of omega-3 fatty acids in psychiatric disorders has been the double-blind, placebo-controlled treatment study conducted by

Stoll and colleagues among subjects with bipolar affective disorder (Stoll et al. 1999b). Bipolar affective disorder is also commonly known as manic–depressive disorder, which more vividly describes the debilitating clinical course of this illness. Thirty subjects were treated with 14 capsules per day containing either 9.6 g/d of ethyl ester EPA plus DHA or an olive-oil placebo. Subjects were studied as outpatients for four months and received the capsules in addition to their regular pharmacological therapies. After four months, there was a significantly reduced risk of relapse to a severe episode of mania or depression in the omega-3-treated group compared to the placebo-treated group. Among subjects taking no other medications, four subjects in the EPA plus DHA group remained symptom free for the length of the study while the four subjects in the placebo group all relapsed. Those results indicate that EPA plus DHA may function as a sole treatment in some patients with bipolar affective disorder. Significant improvements were seen in the Hamilton depression rating scales and the Clinical Global Impressions Scale but not in the Young Mania Depression Scale. These data suggest that depressive symptoms may be more responsive to EPA and DHA than manic symptoms. Further studies in bipolar affective disorder are being conducted by the Stanley Foundation ($n = 240$) and through funding by the Center for Complementary and Alternative Medicine at the National Institutes of Health.

Since the publication of our original hypothesis, only one treatment study of omega-3 fatty acids in major depression has been completed, and none have been published (Marangell et al. 2000). In contrast to the predictions of the tissue-compositional and epidemiological studies, this six-week trial of 2 g/d of DHA alone did not document any differences in depressive symptoms among subjects with mild to moderate major depression. This study of medication-free patients with a Hamilton depression rating scale of greater than 17 and no significant co-morbid psychiatric diagnoses was well designed and carefully conducted. However, several questions remain before the efficacy of DHA in major depression can be ruled out. First, the trial length may not have been adequate. Second, although the dose of DHA (2 g/d) appeared adequate to change most tissue compositions, this dose may have been excessive if there is a non-linear or "inverted U"-shaped pattern of response. In other words, response may be seen at low doses but secondary antagonistic mechanisms may be activated at higher doses. This interpretation is supported by anecdotal reports from some patients who have described better efficacy at doses of 100–300 mg/d (personal observations). Although speculative, it is possible that high doses of dietary DHA may replace EPA or arachidonic acid from small biologically active phospholipid pools (de la Presa-Owens et al. 1998) and decrease eicosanoid production, an effect not apparent at low-dose regimens. Finally, it is also possible that EPA alone or a combination of EPA plus DHA and other essential fatty acids may be needed for optimal treatment efficacy. A greater diversity of controlled clinical studies will help to resolve these questions. Several studies have also described a reduction of depressive symptoms among patients with other primary psychiatric disorders treated with EPA or a combination of EPA plus DHA (Peet et al. 2001, Stoll et al. 1999b). Peet and colleagues reported a 35% decrease in depressive symptoms among schizophrenics treated with 2 g/d of EPA (Peet et al. 2001). Peet and Horrobin (2001) have reported robust treatment responses to 1 g/d of ethyl ester EPA among subjects with treatment resistant depression. Several interventional trials are underway to test these initial findings. These studies do provide suggestive evidence of treatment efficacy for long-chain omega-3 fatty acids in depressive symptoms.

1.6. Cross-national epidemiology of bipolar affective disorder

The cross-national relationships between lower prevalence rates of Bipolar Affective Disorders and greater seafood consumption (Noaghuli, S., Hibbeln, J.R., Weissman, M., unpublished data) are strongly consistent with the clinical intervention trial in bipolar affective disorder described above. Of the psychiatric disorders that have been ex-amined, bipolar spectrum disorders have the strongest, most well-defined relationship to seafood consumption. As with prevalence rates of other psychiatric disorders, there is a non-linear regression relationship. Below an apparent threshold of approxi-mately 75 lbs/person/y, the prevalence rates of bipolar disorder rise precipitously from 0.04% in Taiwan (81.6 lbs/person/y) to 6.5% in Germany (27.6 lbs/person/y), a nearly 60-fold difference in prevalence (Noaghuli, S., Hibbeln, J.R. and Weissman, M., unpub-lished data). Bipolar affective disorders I and II have similar non-linear relationships.

1.7. Postnatal (postpartum) depression

Postpartum, or postnatal, depression may provide an extremely useful model for testing the hypothesis that a deficiency of omega-3 fatty acids in adulthood, and in particular DHA, increases the predisposition to suffering depressive disorders (Hibbeln and Salem 1995). Throughout pregnancy, the placenta actively transfers DHA from the mother to the developing fetus (Campbell et al. 1998). Without adequate dietary replenishment, DHA stores in mothers can become depleted (Al et al. 1994, Holman et al. 1991) and may not be replenished for 26 weeks (Otto et al. 1997, 1999). Given these basic findings, we predicted, and found, that the prevalence rates of postpartum depression would be higher in countries with lower rates of seafood consumption (Hibbeln 1999). The published data on the prevalence of postpartum depression and published data on the DHA content of breast milk was also evaluated (Hibbeln 2001). We found that the DHA concentration of mothers' milk predicted prevalence rates of postpartum depression in a simple linear regression model ($r = -0.88$, $p < 0.0001$, $n = 16$) while seafood consumption predicted prevalence rates of postpartum depression in a non-linear logarithmic regression ($r = -0.81$, $p < 0.0001$). These differences comparing seafood consumption data to the breast-milk compositional data may be due to the non-linear relationship between dietary intake of omega-3 fatty acid and final tissue concentrations (Lands et al. 1992). The observation that the AA or EPA composition of mothers' milk did not predict postpartum depression prevalence rates adds confidence to the proposition that DHA and/or other omega-3 fatty acids are the active components. However, it is important to consider that arachidonic acid may have important psychotropic effects in affective disorders. These cross-national relationships remained significant even after exclusion of all Asian countries and exclusion of countries where there are extreme differences in the percentage of women with low socioeconomic status or other risk factors that have been identified for postpartum depression (Hibbeln 1999). We again note that these cross-national studies are not proof of a causal or protective relationship but they do provide a strong rationale to conduct interventional trials. To our knowledge, there has been no intervention trial completed using either omega-3 or omega-6 fatty acids to prevent or to treat postpartum depression, although two are currently underway at the University of Arizona (Marlene Freeman, M.D., personal communication) and Baylor University in Texas (Lynn Puryor, M.D., personal communication).

2. Impulsive disorders

2.1. Homicide, aggression, hostility and omega-3 status

A review of human intervention studies indicates that it is reasonable to expect that increasing the dietary intake of DHA and EPA might reduce aggression and/or hostility. Virkkunen et al. (1987) reported that violent and impulsive offenders had lower plasma concentrations of DHA and higher concentrations of 22:5 n-6 than non-impulsive offenders and healthy controls. Three double-blind, placebo-controlled intervention trials have demonstrated the efficacy of omega-3 fatty acids in reducing hostility, an affective state closely related to anger and aggression. One double-blind, placebo-controlled trial was specifically conducted to assess the efficacy of DHA in reducing measures of hostility. Hamazaki et al. (1996) reported that 1.5 to 1.8 g/d of DHA reduced measures of hostility in a picture frustration test among Japanese students undergoing the stress of University exams, in comparison to a placebo-treated group. One significant aspect of this study is that the baseline plasma DHA composition of this group was 3.0% (compared to typical American levels of less than 1%). Hostility was assessed with the Rosensweig Picture Frustration Test. Thienprasert et al. (2000) also reported decreases in hostility measures among Thai university employees, but not rural farmers in a 2-month double-blind trial of 1.5 g/d of DHA. Consistent with these reports, a five-year intervention trial documented reductions in hostility and depression scores among subjects consuming a high-fish diet (Weidner et al. 1992). Thus, it is interesting that across 26 countries, homicide mortality rates are correlated with lower rates of seafood consumption ($r = 0.63$, $p < 0.0006$) over a 30-fold difference in prevalence (Hibbeln 2000). This cross-national study did not demonstrate that increasing omega-3 consumption caused lower rates of homicide. That assertion is best tested in a controlled interventional trial.

2.2. Suicide and impulsiveness

Impulsiveness is defined as the inability to inhibit behavior and has frequently been proposed as an important deficit among aggressive and suicidal subjects. Many theorists have suggested that many aggressive behaviors are a subset of impulsive behaviors (Apter et al. 1993, Mann 1998). Mann (1998) has also suggested that aggression and suicidality may also have a common behavioral denominator, impulsiveness, which is associated with a common biological correlate, low serotonergic function. The putative relationships between low omega-3 status and serotonergic function will be discussed below. We note that no interventional trials have been conducted to test if improving omega-3 status reduces impulsiveness. However, several studies indicate a low omega-3 status is associated with greater impulsiveness:

(1) In a study of 265 000 Japanese men followed for 17 years, Hirayama (1990) found that daily fish consumption significantly reduced the risk of death due to suicide (odds ratio 0.81) compared to eating fish less than every day.

(2) Among 1434 normal subjects in Northern Finland, frequent fish consumption was associated with a reduced risk of suicidal thinking (odds ratio 0.57) (Tanskanen et al. 2001a).

(3) Among 165 subjects with major affective disorders, lower plasma DHA and EPA concentrations predicted higher scores on the Barratt Impulsiveness Scale (Noaghiul et al., unpublished).

(4) Among impulsive suicide attempters, low concentrations of plasma EPA predicted higher ratings of impulsiveness, greater guilt, higher future suicide risk and greater psychopathology including depression (Hibbeln et al. 2000c).

These data appear sufficient to raise the possibility that a low omega-3 status may be associated with a greater risk of suicide.

2.3. Attention deficit hyperactivity disorder

Initial reports of deficiencies of essential fatty acids among children with attention deficit hyperactivity disorder (ADHD) and dyslexia raised the promise of potential non-drug treatments. However, convincing treatment data from interventional trials has not yet been reported. Stevens et al. (1995) found that 53 subjects with attention deficit hyperactivity disorder had significantly lower concentrations of AA, EPA, and DHA in plasma polar phospholipids when compared to 43 control subjects. One investigator (Stordy 1995) described decreased rod function in 10 young dyslexics compared to 10 controls. In an open trial, supplementation with a fish oil containing 480 mg of DHA/d for one month improved scotopic vision among these dyslexics. Stordy also reported that supplementation with a mixture of essential fatty acids improved motor skills in a open trial of 15 ADHD children (Stordy 2000). Two well-controlled, double-blind placebo-controlled trials among children with attention deficit hyperactivity disorder supplemented with either DHA (Voigt et al. 1998) or a mixture of EPA plus DHA (Burgess 1998) were presented at an international workshop in 1998. Neither study found positive treatment effects. Richardson et al. (personal communication) have reported decreased ADHD symptoms among children with speech disorder. However, to our knowledge, neither positive nor negative results from these studies have appeared in publication. One important confounding factor is that the diagnosis of behavioral disorders among children is often very difficult. For example, many children with bipolar affective disorder may respond to supplementation but be mistaken for children with attention deficit hyperactivity disorder. The published data are insufficient to either accept or reject the hypothesis that omega-3 supplementation is useful in treating behavioral disorders among children such as attention deficit hyperactivity disorder. Great attention should be paid to the observation that childhood affective disorders are very difficult to distinguish from attention deficit hyperactivity disorder (Giedd 2000). These may actually be improvements in early affective symptoms. It is important to continue to listen to the anecdotal reports of parents and teachers who have seen dramatic improvements in disruptive behaviors in order to reduce our ignorance regarding the diagnoses of these children.

2.4. Alcoholism

A low omega-3 status among alcoholics may in part explain the high rates of depression and violence among this population. Depression secondary to alcoholism is common as it occurs in 16% to 59% of alcoholic patients (Merikangas and Gelertner 1990, Winokur 1990) who represent 5–10% of the US population. Depression occurs more frequently in alcoholics (58%) than in opiate addicts (32%) or schizophrenics (28%) (Weissman 1983). Schuckit (1986) noted serious depression in 70% of patients with prolonged heavy drinking and argued that these depressions were the consequence of the pharmacological effects of alcohol intoxication, withdrawal and life crises. Severe depressive symptoms remitted within 4–6 weeks of abstinence alone. At high levels of intake, alcohol is a pro-oxidant that leads to increased lipid peroxidation (Bjorneboe and Bjorneboe 1993, Rosenblum et al. 1989). A consequence of increased lipid peroxidation may be a decrease in the levels of the more highly unsaturated species such as 22:6 n-3. Several studies have demonstrated that chronic alcohol intoxication depletes long-chain n-3 polyunsaturated fatty acids from neuronal membranes (Salem and Ward 1993) which we suggest may facilitate development of depressive symptoms. Chronic alcohol consumption also markedly depleted DHA from neuronal membranes (Pawlosky and Salem 1995, 1998) among adult rhesus monkeys that drank alcohol at moderate levels. These animals had been given a diet that contained low, but adequate, concentrations of essential fatty acids. The fatty acid composition of these diets resembled diets of some American alcoholics. Chronic alcohol consumption also decreases central nervous system serotonin concentrations (Higley and Bennett 1999), which may be responsible for increased impulsivity. The decreased central nervous system concentrations of serotonin caused by alcohol may be reversed by a diet with adequate amounts of essential fatty acids (Olsson et al. 1998). Both chronic alcohol consumption and a diet containing low, but adequate amounts of essential fatty acids, lowered serotonin and dopamine concentrations in rat cortex to similar levels, when compared to a diet containing just 0.5% AA and 0.5% DHA (Olsson et al. 1998). Thus, we postulate that increasing dietary intake of DHA plus EPA among abstinent alcoholics will restore central and peripheral omega-3 status. This restoration of omega-3 status may increase brain serotonin concentrations and reduce impulsive behaviors among these alcoholics.

3. Possible mechanisms: neurotransmitter function

Abnormalities in serotonergic function are thought to be important in impulsive, suicidal and depressive behaviors. One of the best-replicated findings in biological psychiatry is that low concentrations of 5-hydroxyindoleacetic acid in cerebrospinal fluid (CSF 5-HIAA, a metabolite of serotonin) are associated with suicide and depression (Roy et al. 1987). Low CSF 5-HIAA concentrations predict impulsive, hostile and aggressive behaviors (Linnoila et al. 1983, 1989, Mann et al. 1999, Virkkunen et al. 1994) and reflect serotonin turnover in the frontal cortex (Stanley et al. 1985).

One simple core clinical question is: does increasing the dietary intake of omega-3 fats improve concentrations of neurotransmitters that are implicated in depressive and violent behaviors? Therefore, it is important to examine the animal literature that documents that altering the omega-3 fatty acids composition of an animal's diet can alter neurotransmitter concentrations. It is also important to note that deficiencies in omega-3 fatty acids impact monoaminergic neurotransmission in the frontal cortex, which has been shown to regulate both impulsive behaviors and cardiovascular reactivity in response to emotional stress. Zimmer and colleagues (2000b) documented a significant reduction in the number of dopaminergic synaptic vesicles in the frontal cortex of omega-3-deficient rats. This dietary deficiency in omega-3 fatty acids resulted in a 90% reduction in the quantity of dopamine released after tyramine stimulation (Zimmer et al. 2000a). In comparing diets containing fish oils to diets deficient in omega-3 fats, others (Chalon et al. 1998) found that dopamine levels were 40% greater in the frontal cortex of rats fed fish oils compared to those fed the control diet. In frontal cortex there was also a reduction in monoamine oxidase activity and greater binding to dopamine D_2 receptors.

While animal studies have most strongly documented improvements in dopaminergic function, they also indicated that the function of serotonergic neurons may also be altered in adult animals. DeLion and colleagues (1996) reported that a chronic dietary deficiency in omega-3 fatty acids specifically affected monoaminergic systems in the frontal cortex of rats. They reported that a 40–75% lower level of endogenous dopamine in the frontal cortex occurred in deficient rats according to age. This deficiency also induced a 10% reduction in the density of dopaminergic D_2 receptors and an 18–46% increase in serotonin 5-HT_2 receptor density in the frontal cortex with no change in binding affinity and without variation in serotonin levels. These changes in serotonin 5-HT_2 receptor density were strikingly similar to the abnormalities noted by Stanley and colleagues (1983) among victims of suicide; there was a 44% increase in 5-HT_2 receptor density and no change in binding affinity in the frontal cortex. Heron and colleagues (1980) found that changing the fatty acid composition altered membrane biophysical properties that resulted in markedly altered serotonin receptor binding. These observations support the assertion that greater omega-3 fatty acid intake may improve the abnormalities of serotonergic neurotransmission, which are associated with impulsive behaviors.

3.1. Neurotransmitter function in non-human infants

Neurotransmission in the frontal cortex can also be affected by dietary essential fatty acids during infancy. De la Presa Owens and Innis (1999) fed piglets one of four infant formulas for 18 days. The formulas were either adequate (C−) or deficient (D−) in 18:2 n-6 and 18:3 n-3 or contained supplemental arachidonic acid (AA) (0.2%) and DHA (0.16%), D+ and C+, respectively. Frontal-cortex concentrations of serotonin, tryptophan, dopamine, HVA and norepinephrine were nearly doubled in the D+ and C+ formulas. It is remarkable that only 18 days of dietary intervention altered concentrations of these fundamental neurotransmitters. Austead and colleagues (2000) also reported changes in concentrations of both serotonin and CSF 5-HIAA in frontal cortex among piglets given control and DHA/AA supplemented formulas. They reported that frontal-cortex concentrations of serotonin increased while the concentration of the metabolite, CSF 5-HIAA, decreased. One interpretation of these data is that DHA/AA supplementation increases tissue concentrations of serotonin by decreasing degradation of serotonin to its metabolite,

CSF 5-HIAA (personal communication). Several drugs used to treat depression have similar effects of raising tissue concentrations of serotonin and lowering CSF 5-HIAA concentrations. For example, fluoxetine (Prozac®) increases the concentration of serotonin in the synapse by inhibiting reuptake into the cell (Paez and Hernandez 1998). CSF 5-HIAA decreases because the serotonin is not exposed to the degradative enzyme monoamine oxidase (MAO). Consistent with this interpretation, DeLion and colleagues (1997) reported that diets rich in omega-3 fats lead to a decrease in MAO activity.

Among infant rhesus monkeys, we found that supplementing their infant formula with AA (1.0%) and DHA (1.0%) (DHA/AA group) improved neurodevelopmental outcomes and heart-rate variability, but decreased CSF 5-HIAA concentrations (Hibbeln et al. 2000b). In this experiment, infants were removed from their mothers at birth and, for six months, received one of the two formulas while being raised in a stringently controlled nursery. The DHA/AA group received formulas supplemented with AA (1.0%) and DHA (1.0%) which is similar to the milk of rhesus monkey mothers. The control formulas were similar to commercially produced human infant formulas as they were virtually devoid of DHA and arachidonic acid (AA). At six months all animals were switched to diets rich in omega-3 fatty acids. DHA/AA infants had profoundly improved motor development and visual orientation scores in as little as seven days (Champoux et al. 2001). As described above, heart-rate variability remained improved in adolescence, up to 3.5 years after the dietary intervention had stopped, indicating an enduring developmental effect. CSF 5-HIAA was decreased in the DHA/AA group, but only during the six months of formula feeding. We cannot determine directly whether the supplementation raised or lowered the brain concentrations of serotonin among these infants. However, the behavioral and physiological improvements noted above, were consistent with improved serotonergic function.

3.2. Human data on omega-3's and neurotransmitter metabolites

Correlational data from human studies are consistent with the proposition that omega-3 status is related to CSF neurotransmitter metabolite concentrations. We observed that plasma concentrations of DHA and AA predicted CSF 5-HIAA and HVA concentrations in 234 subjects investigated at the National Institute on Alcohol Abuse and Alcoholism (Hibbeln et al. 1998a,b). In healthy control subjects and late-onset alcoholics, higher concentrations of plasma DHA predicted higher concentrations of CSF 5-HIAA. It is remarkable that this correlational relationship was found between a cerebrospinal fluid measure of a neurotransmitter metabolite and a plasma level of any fatty acid. We have also replicated this finding among 104 adult rhesus monkeys. Higher concentrations of the omega-3 fatty acids DHA and EPA in plasma predicted higher concentrations of CSF 5-HIAA (Hibbeln, Higley et al., unpublished). Among these animals, higher EPA and DHA plasma concentrations also predicted more functional dominance behaviors. We note that these animals were all on the same diet so their variance in plasma fatty acid compositions were due to individual differences in fatty acid metabolism, catabolism or responses to the fatty acids. These correlational findings do strongly suggest that increasing DHA intake may increase brain serotonin concentrations. Higher brain concentrations of CSF 5-HIAA may reduce depression and aggression. Direct data demonstrating that increasing EPA or DHA status will change CSF 5-HIAA concentrations among human subjects are sparse. Nizzo and colleagues (1978) reported that acute intravenous infusions of DHA-containing phospholipids increased CSF 5-HIAA and homovanillic acid (HVA) concentrations in human subjects over six hours. Unfortunately this study was poorly controlled and examined few subjects.

3.3. Deficiencies in gestation or early development and psychiatric outcomes

An intriguing possibility is that some level of deficiency in omega-3 status either during interuterine, postnatal or in later development, could contribute to a lifelong risk of suffering psychiatric illnesses through irreversible neurodevelopmental changes, as has been proposed for schizophrenia (Peet et al. 1999). For example, studies of periods of famine such as the Dutch Hunger Winter have generated hypotheses that specific psychiatric illness may be related to nutritional deprivation in specific periods of development. Prenatal malnutrition in the first trimester may increase risk for developing schizophrenia (Susser and Lin 1992) as reflected in increased sulcal sizes (Hulshoff Pol et al. 2000) whereas later gestational famine during the 2nd and 3rd trimester increased the risk for later affective disorders (Brown et al. 2000). One test of the hypothesis that mothers of children with psychoses are deficient in essential fatty acids during gestation has been to examine the composition of maternal plasma on the day of birth and the development of psychotic illnesses over the next four decades among their children (Hibbeln et al. 2000a). The total fatty acid content as well as the cholesterol ester 18:3 n-3 and 18:2 n-6 were significantly higher in the plasma from mothers of 27 children that developed psychosis compared to the plasma of the 51 control mothers matched for age, date of birth and ethnicity. Cholesterol ester DHA and cholesterol ester EPA concentrations were also elevated, but were not statistically different. These data are inconsistent with the hypothesis that mothers of children with psychoses are deficient in n-6 or n-3 essential fatty acids.

It is also possible that essential fatty acid deficiencies in the postnatal period may increase the risk of the development of psychiatric or behavioral disorders. Higley and colleagues (1991) have extensively validated a model for inducing greater risk for lifetime behavioral disturbances and alcohol preference among rhesus monkeys. In this paradigm, infant rhesus monkeys were separated from their mothers at birth, raised in a nursery with a cloth surrogate mother and fed standard formulas. These infant formulas resembled human infant formulas commercially available in the USA in that they are virtually devoid of DHA and AA (Hibbeln et al. 2000b). Several studies have documented that infants fed standard formulas virtually devoid of DHA and AA have sub-optimal neurological development (as evidenced by behavior and visual outcome measures), when compared to infants fed formulas supplemented with DHA and AA (Agostoni et al. 1995, Birch et al. 2000, Carlson et al. 1991, Gibson and Makrides 2000, Willatts et al. 1998). Thus, we postulated that some portion of the increased risk for aggressive and depressive behaviors among the nursery-raised rhesus monkeys described above could be due to the very low levels of dietary AA and DHA. It is known that maternal–infant interactions have profound effects on early development and maturation so a nutrition deprivation could at best contribute to only a small portion of the behavioral differences observed when mother-raised and nursery-raised infants are compared. Long-term data on behavioral differences among these primates are still being assessed. It will be difficult to design human studies that adequately test the hypothesis that formulas virtually devoid of AA and DHA increase the predisposition to violence or aggression.

4. Cardiovascular disease and depression: introduction

Major depression and cardiovascular disease are debilitating and widespread disorders. The Seven Countries Study on Cardiovascular Diseases found that 26.3% of deaths were due to cardiovascular mortality (Menotti et al. 1989), and the World Health Organization estimated in 1996 that major depression was the single greatest cause of years of life lost to disability worldwide (Murray and Lopez 1996). Therefore, it is of great clinical significance that over the past 30 years, both depressive symptoms and hostility have been established as major risk factors for cardiovascular mortality in prospective and retrospective studies (Booth-Kewley and Friedman 1987, Carney et al. 1987, Ford et al. 1998, Glassman and Shapiro 1998, Kaplan et al. 1991b, Musselman et al. 1998). Several factors have been identified as risk factors of cardiovascular morbidity among depressed patients. These include an increased risk of sudden cardiac death, increased risk of cardiac arrhythmias, heightened platelet aggregation and activation, decreased heart-rate variability, elevated plasma norepinephrine, and hyperactive hypothalamic–pituitary–adrenocortical axis activity. Several authors have postulated that depressed states elevate sympathetic tone which, in turn, may increase vascular and platelet reactivity, create alterations in blood pressure and cause endothelial injury, thus increasing cardiovascular risk (Carney et al. 1987, Ford et al. 1998, Musselman et al. 1998). Finally, it has been proposed that elevations in corticotrophin releasing factor (CRF), which then stimulates increases in sympathoadrenal and HPA-axis activity, is a primary mechanism underlying these risk factors (Musselman et al. 1998). However, no hypothesis yet proposed provides a causal explanation of the origin of elevated sympathetic tone among depressed patients and describes specific interactions with each risk factor. We postulate that a low or insufficient tissue status of EPA and DHA exacerbates each of the cardiovascular risk factors that have been described among depressed patients. Although our original statement of our hypothesis that EPA and or DHA insufficiency may link depression to CVD (Hibbeln and Salem 1995) has been followed by a brief discussion of possible mechanisms (Severus et al. 1999), this proposition merits a more careful examination.

4.1. Omega-3 fatty acids in cardiovascular disease (CVD)

Several reviews have already described the protective roles of omega-3 fatty acids in reducing the morbidity and mortality of cardiovascular disease (Abeywardena et al. 1992, Leaf and Weber 1987). A few outstanding studies will serve to clarify the importance of EPA and DHA in cardiovascular risk. In a treatment trial involving 11 324 post-infarction subjects, one gram of EPA plus DHA per day significantly lowered the risk of subsequent cardiovascular death by 30%, and all-cause death by 20% [GISSI 1999].

In the MRFIT study of 12 866 men, the estimated sum of EPA, docosapentaenoic acid (n-3) and DHA consumption was negatively correlated with cardiovascular mortality after adjustment for cardiovascular risk factors (ARR = 0.59, $p < 0.005$) (Dolecek and Granditis 1991). Finally, in three different studies of populations with low dietary fish intake, fish consumption was negatively correlated with cardiovascular mortality (Kromhout 1989). Even a slight increase in dietary intake of LNA was associated with a decreased relative risk of myocardial infarction of 0.41 ($p < 0.01$) following adjustment for risk factors (Ascherio et al. 1996). These and other studies have demonstrated the validity of the protective relationship of omega-3 fats for cardiovascular diseases.

4.2. Mechanisms of increased cardiovascular risk in depression

Six primary factors have been identified among depressed patients that increase risk of cardiovascular death (Musselman et al. 1998). Each of these factors is exacerbated by low omega-3 status. These factors are increased risk of sudden cardiac death, increased vulnerability to cardiac arrhythmias, increased platelet activity, decreased heart-rate variability, elevated plasma norepinephrine, and overactivity of the hypothalamic–pituitary–adrenocortical (HPA) axis. CRF-induced increases in both sympathoadrenal and HPA-axis activity have been previously proposed as the primary mechanism underlying these risk factors in depression (Musselman et al. 1998). We will examine both the effects of omega-3 fats in reducing sympathoadrenal and HPA-axis hyperactivity and the independent protective effects on each factor. The relationship of omega-3 fatty acids to elevated triglycerides, homocysteine levels and serotonergic function in depression and CVD will also be discussed, although the evidence supporting these relationships is not fully developed. Each of these factors will be discussed in turn, with respect to their relationships with CVD, depression, and omega-3 status.

4.2.1. Sudden cardiac death

Sudden cardiac death accounts for at least half of cardiac deaths (Deshpande and Akhtar 1997, Myerburg et al. 1997) and is the most common form of death for men between the ages of 25 and 65 (Reich 1985). The association between sudden cardiac death and depression was first noted in the early 1970s, when two separate investigators noted that depressive symptoms often predated sudden cardiac death (Bruhn et al. 1974, Greene et al. 1972). Later studies have confirmed that depression can predict sudden cardiac death in post-infarction patients (Ahern et al. 1990, Frasure-Smith et al. 1995, Ladwig et al. 1991), particularly when the depression is untreated (Goldstein and Niaura 1992, Reich 1985). Increased fish consumption or elevated omega-3 status clearly reduces risk of sudden cardiac death. Fish intake equivalent to 1 fatty fish per week was associated with a 50% reduction in risk of primary cardiac arrest in 334 patients with primary cardiac arrest and 493 controls (Siscovick et al. 1995). In 20 551 subjects originally free from myocardial infarct, fish intake at least once per week was again associated with a greater than 50% reduction in risk of sudden cardiac death (RR = 0.48, $p < 0.04$) (Albert et al. 1998). The cellular mechanisms by which omega-3 fats reduce risk of sudden cardiac death are becoming increasingly evident. In the canine model of sudden cardiac death, EPA and DHA administered as intravenous free fatty acids reversed ongoing arrhythmias induced by coronary artery occlusion (Kang and Leaf 1996, Leaf and Kang 1996). Thus, the low omega-3 status of depressed patients may be responsible for their increased risk of sudden cardiac death.

4.2.2. Arrhythmias

Myocardial infarctions frequently increase the electrical instability of cardiac tissue and initiate arrhythmias, the inability of cardiac muscles to contract in a coordinated manner. This inability of the heart to function properly can rapidly lead to sudden cardiac death, and accounted for 84% of occurrences of sudden cardiac death across 7 published series (Bayes de Luna et al. 1989). Depressive symptoms appear to increase the risk for and severity of arrhythmic events. In a study of 103 patients with pre-existing CVD, patients with either major or minor depression were more likely ($p < 0.01$) to have episodes of ventricular tachycardia than non-depressed controls after adjustment for co-variates (Carney et al. 1993). In 218 post-infarct patients, those with comorbid depression and arrhythmic premature ventricular contractions were 29.17 times as likely to die from cardiac causes in the 18 months following infarct compared to all other patients ($p < 0.00001$). Five of these six cardiac deaths were arrhythmogenic (Frasure-Smith et al. 1995).

Animal and cellular studies have helped to elucidate mechanisms by which omega-3 fats abate cardiac arrhythmias. Elevated myocardial EPA and DHA from dietary sources prevent the initiation and reduce the severity of arrhythmias in isolated rat hearts in response to ischemia, reperfusion, and programmed electrical arrhythmias (Hsu and Yap 1967). Dietary supplementation with small amounts of DHA has also been shown to inhibit ischemia-induced cardiac arrhythmias *in vivo* in rats (Pepe and McLennan 1996). EPA and DHA appear to exert these antiarrhythmic effects via sodium-channel-mediated electrophysiological modifications (Abeywardena et al. 1992, Leaf and Kang 1996), the inhibition of L-type calcium channels (Hallaq et al. 1992), and modulation of eicosanoid production (Abeywardena et al. 1992).

4.2.3. Heart-rate variability (HRV)

Despite the complexity involved in the interpretation of heart-rate variability (HRV) data, low HRV has been strongly associated with post-infarct mortality, depression, and insufficient EPA and DHA intake. HRV is measured in Hertz, and is generally separated into low-, middle-, and high-frequency (or power) domains. Although there has been some disparity between researchers with regards to the precise boundaries of each of these domains, there has been considerable consistency in reporting that low-power HRV is under the influence of both the sympathetic and parasympathetic nervous systems (Akselrod et al. 1985, Koizumi et al. 1985, Liao et al. 1997), whereas middle- and high-power HRV appear to be solely under parasympathetic influence (Akselrod et al. 1985, Liao et al. 1997). In addition, the assumption that vagal tone is nearly synonymous with myocardial parasympathetic tone appears to be valid (Bailey et al. 1996).

The predictive power of low HRV for subsequent mortality in patients following myocardial infarction has been well replicated (Bigger et al. 1988, 1993, Billman et al. 1982, Farrell et al. 1991, Kleiger et al. 1987, Odemuyiwa et al. 1994), and has been shown to retain its validity independently of other risk factors. Lowered HRV has also been associated not only with major depressive disorder but also with minor depression and depression rating in non-depressed subjects (Carney et al. 1995, Krittayaphong et al. 1997, Light et al. 1998, Tulen et al. 1996). Although it has been suggested that HRV in depressed patients improved with successful antidepressant treatment (Balogh et al. 1993, Khaykin et al. 1998), these findings have been contested (Yeragani et al. 1992). It is interesting to note here that stimulation of the vagus nerve with implantable electrodes has been successfully utilized in the treatment of depressive subjects resistant to antidepressant medications and to electroconvulsive therapy (Rush et al. 2000).

Higher concentrations of DHA and EPA in platelets predicted increased HRV both at baseline and after dietary intervention, despite HRV quantification using a relatively crude measure, the standard deviation of RR intervals. At baseline, greater HRV was positively correlated to DHA content, and the DHA to AA ratio in platelets among post-infarction patients (Christensen et al. 1997). Following twelve weeks of omega-3 supplementation, these same patients showed increases in platelet omega-3 content and substantial increases in HRV relative to both baseline measures and control subjects (Christensen et al. 1996). In a separate study, twenty-nine patients with chronic renal failure were treated with 5.2 g of EPA and DHA or a placebo for 12 weeks while undergoing dialysis (Christensen et al. 1998). The highly significant positive correlation between omega-3 content in granulocyte membranes and HRV ($r = 0.71$, $p < 0.01$) and the significant clinical treatment effect compared to placebo indicated improvement in cardiovascular autonomic dysfunction among these subjects. Among male healthy volunteers, HRV improved in a dose-dependent manner after omega-3 supplementation (Christensen et al. 1999) although no such effect was found in women. Hence, the findings of decreased HRV among depressed patients may be related to their low EPA and DHA status. However, we found no difference in HRV between third-generation rats raised on either an omega-3-enhanced or omega-3-deficient diet, after analysis with the Hurst exponent, a more sophisticated measure of fractal systems (Weisinger et al., unpublished).

4.2.4. Platelet reactivity

Platelet activation due to interaction with the arterial endothelium is heightened in depressed patients, is attenuated by EPA and DHA, and plays a key role in the formation of coronary arterial thrombi. Thrombus formation causes acute cardiovascular syndromes including unstable angina pectoris, acute myocardial infarction and sudden cardiac death (Kristensen et al. 1989). Initiation of platelet activation can occur as a result of damage to the endothelial barrier, which leads directly to the adhesion of platelets as well as to the release of pro-aggregatory and vasoactive substances such as Thromboxane A_2 (TxA_2), Adenosine diphosphate (ADP), and serotonin. Although serotonin is but a weak platelet activator on its own, it is able to elicit strong platelet reactions through amplification via arachidonic acid and TXA_2 release (De Clerck 1991). This amplification occurs when serotonin stimulates 5-HT_2 receptors and induces turnover of phosphatidylinositol 4,5-bisphosphate and production of D-*myo*-inositol 1,4,5-trisphosphate (Ins(1,4,5)P_3). Ins(1,4,5)P_3 releases calcium from intracellular stores, thereby activating phospholipases (including phospholipase A_2), which release fatty acids from membrane (De Clerck 1991, Kristensen et al. 1989). These fatty acids are then available for conversion to eicosanoids. Eicosanoids derived from arachidonic-acid release are pro-aggregatory and pro-inflammatory, while eicosanoids derived from EPA weakly inhibit aggregation and inflammation.

Two key regulators of platelet–vessel wall interactions are the eicosanoids TxA_2 and prostacyclin (PGI_2). Platelets are the primary source of TxA_2, which has platelet pro-aggregatory and vasoconstrictive properties. PGI_2, in contrast, is formed mainly by epithelial cells, causes vasodilation, and inhibits platelet aggregation. After reviewing the effects of omega-3 fatty acids on platelet function and bleeding time in 37 human feeding trials, Kristensen and colleagues (1989) concluded that supplementation with EPA and DHA produced a favorable shift in the PGI_2/TXA_2 balance and reduced platelet aggregation and vasoconstriction. DHA has also been shown to reduce platelet reactivity independently of its conversion to EPA (von Schacky and Weber 1985). Healthy men given supplements of EPA and DHA for four months showed decreases in platelet activity, which remained detectable after two months of washout (Prisco et al. 1995). In addition to inhibiting acute thrombotic reactions, EPA and DHA virtually abolish smooth muscle cell proliferation and intimal hyperplasia, which are normally stimulated after endothelial damage and are key pathological features underlying the development of atherosclerotic lesions (Pakala et al. 1999b,c).

Heightened platelet activation has been described in depressed patients compared to ischemic and non-ischemic controls (Laghrissi-Thode et al. 1997, Musselman et al. 1996). Twelve medication-free depressed patients showed signs of heightened coagulant activity at baseline ($p < 0.05$) as well as a greater number of significant increases in measures of platelet reactivity to orthostatic challenge compared to eight well-matched controls (Musselman et al. 1996). After considering the important and well-documented roles of EPA and DHA and of eicosanoids derived from EPA in reducing platelet aggregation and in mediating endothelial wall interactions, it is reasonable to consider that abnormalities in platelet aggregation found among depressed patients may in part be explained by insufficient tissue concentrations of these fatty acids.

4.2.5. HPA-axis hyperactivity: immune-neuro-endocrine mechanisms

Substantial evidence suggests that hyperactivity of the hypothalamic–pituitary–adreno-cortical (HPA) axis has a central role in increasing vulnerability to cardiovascular disease among depressed patients (Musselman et al. 1998). This HPA-axis hyperactivity appears to be driven by the elevated concentrations of corticotrophin releasing factor (CRF) found in the cerebrospinal fluid and in hypothalamic neurons of depressed patients (Banki et al. 1992, France et al. 1988, Purba et al. 1995, Raadsheer et al. 1995). A considerable body of data has documented that elevations in CRF concentrations and hyperactivity of the HPA axis can be stimulated by the release of cytokines by macrophages, in particular the cytokines interleukin-1β (IL-1β), interleukin-6 (IL-6) and tumor necrosis factor (TNF) (see the review by Turnbull and Rivier 1999). Elevated IL-1β concentrations also positively correlate with post-dexamethasone suppression test cortisol values in both depressed and healthy subjects, suggesting that elevated IL-1β concentrations contribute to HPA-axis hyperactivity (Maes et al. 1993a). Consistent with these data, elevated IL-1β and IL-6 concentrations have been repeatedly described among depressed patients (Maes et al. 1991, 1993a,b, 1995a,b, Sluzewska et al. 1996) as well as in patients with ischemic heart disease and congestive heart symptoms (Hasdai et al. 1996, Testa et al. 1996). In addition, IL-6 and TNF appear to increase cardiovascular risk factors (Jovinge et al. 1998, Mendall et al. 1997). Thus, factors that promote excessive cytokine release may contribute to HPA-axis hyperactivity among patients with CVD and depression.

R.S. Smith (1991) first suggested that HPA-axis hyperactivity may be stimulated by an insufficient omega-3 status by allowing excessive IL-1β release by macrophages and elevating CRF concentrations. In concordance with that hypothesis, dietary omega-3 supplements have been shown to suppress the synthesis of IL-1β, IL-6, and TNF in human subjects (Gallai et al. 1995, Kremer et al. 1990, Meydani et al. 1993). Stimulated IL-1β production in normal subjects was reduced by 40% after adding EPA and DHA to National Cholesterol Education Panel step-2 diets (Meydani et al. 1993). In another set of normal subjects, IL-1β production was suppressed by 43% after supplementation with fish oil (Endres et al. 1989). IL-1 production was reduced by 54% and clinical improvements were noted among patients with rheumatoid arthritis following supplementation with EPA and DHA (Espersen et al. 1992, Kremer et al. 1990). Peripheral cytokines act across the blood–brain barrier by stimulating eicosanoids such as prostaglandin E2 (PGE2), which in turn activates the central release of IL-1β and CRF in the central nervous system (Turnbull and Rivier 1999). In fact, the synthesis of PGE_2 in the central nervous system was necessary for interleukin-induced production of CRF from the rat hypothalamus (Navarra et al. 1992) as well as for corresponding changes in blood pressure, heart rate, and renal sympathetic nerve activity (Kannan et al. 1996). Omega-3 fats clearly alleviate excessive cytokine release, as demonstrated by the large clinical responses to supplementation. Given that the central nervous system production of CRF in response to peripheral IL-1β also depends upon eicosanoid generation, it seems likely that EPA and DHA may moderate the cytokine-driven component of HPA-axis hyperactivity associated with increased cardiovascular risk in depression.

4.2.6. Eicosanoids mediate CNS-driven sympatho-adrenal and HPA-axis hyperactivity

Elevated concentrations of hypothalamic corticotrophin releasing factor (CRF) have been proposed as an important drive of both sympathetic hyperactivity (Musselman et al. 1998) and HPA-axis hyperactivity (Banki et al. 1992, France et al. 1988, Purba et al. 1995, Raadsheer et al. 1995). We propose that the excessive release of AA-derived eicosanoids in the central nervous system (CNS) may contribute to excessive sympathetic and HPA activity driven by CRF. In addition to modulating the release of cytokines in the periphery, the capacity of CNS cytokines to stimulate the sympathetic nervous system activity (Kannan et al. 1996, Murakami et al. 1996, Ohashi and Saigusa 1997) and to inhibit the parasympathetic vagal activity (Nair et al. 1997) depends upon prostaglandin synthesis. The ability of cytokines and noradrenergic pathways to stimulate CRF release in the hypothalamus is also dependent upon the production of prostanoids derived from AA, such as PGE_2 and $PGF_{2\alpha}$ (Bugajski 1996, Terao et al. 1993). Diets high in fish oil (14% cod liver oil or 12.5% salmon oil) can reduce brain concentrations of AA, 22:4 n-6 and 22:5 n-6, and raise brain concentrations of DHA, as was described in comparison to 8 weeks of a corn oil diet (Bourre et al. 1988). Similar decreases in neuronal AA concentrations have been seen during early development (Philbrick et al. 1987) and during recovery from a multigenerational omega-3 deprivation (Homayoun et al. 1988). Thus, an insufficient omega-3 status may allow for excessive release of AA-derived eicosanoids in hypothalamic neurons which, in turn, stimulates excessive CRF release and sympathetic and HPA axis hyperactivity.

4.2.7. Plasma norepinephrine (NE)

NE release has been well documented to precipitate arrhythmias in the ischemic myocardium (Schomig et al. 1995). *In vivo*, elevated plasma NE was the strongest predictor of cardiovascular mortality (RR = 2.55) and myocardial ischemia (RR = 2.59) in 514 patients with left ventricular dysfunction (Benedict et al. 1996). Chronic elevations in plasma NE are thought to contribute to the development of cardiovascular disease as well, through direct chronotrophic effects of NE on the heart and adrenergic-mediated vasoconstriction. Resulting alterations in hemodynamic and shear forces may cause damage to the epithelial layer of arteries and initiate development of atherosclerotic lesions (Kaplan et al. 1991b). Factors that slow this progression include stress reduction and diminished sympathetic hyperactivity as well as protection of the tissues from the effects of elevated plasma NE concentrations with the use of beta-adrenergic blocking agents (Kaplan et al. 1991b).

Elevated plasma NE concentrations have also been linked to major depression (Lechin et al. 1995, Rudorfer et al. 1985, Veith et al. 1994) as well as depressive symptoms in both healthy (Light et al. 1998) and depressed (Lechin et al. 1995) subjects. One study, however, found no differences between depressed and non-depressed patients with cardiovascular disease with regard to plasma NE (Carney et al. 1999). It is interesting to note that elevations in plasma NE among depressed patients are most likely due to sympathetic hyperactivity (Fielding 1991) and that stress-induced increases in plasma NE are more prominent in healthy subjects with high depression scores (Light et al. 1998).

4.3. A central role of excessive sympathetic tone

Each of the six primary mechanisms proposed to contribute to the increased cardio-vascular risk associated with depression is aggravated by sympathetic nervous system hyperactivity. Increased sympathetic tone has been shown to increase risk of sudden cardiac death (Barron and Lesh 1996, Schwartz et al. 1992, Willich et al. 1993) whereas increased parasympathetic tone to the heart decreases risk of sudden cardiac death (Barron and Lesh 1996, Schwartz et al. 1992). Sympathetic activity resulting from direct cortical stimulation was able to induce arrhythmias in dogs (Hockman et al. 1966). Elevated sympathetic activity also decreases heart-rate variability (Bigger et al. 1988, Farrell et al. 1991, Odemuyiwa et al. 1994). The increases in platelet activation documented in depressed patients occurred in response to a stimulator of sympathetic activity (Musselman et al. 1996) which modifies platelet activity through alterations in vascular eicosanoid synthesis (Anfossi and Trovati 1996). Since plasma NE derives primarily from sympathetic nerve terminals (Kopin et al. 1978, Musselman et al. 1998), it is therefore indicative of both sympathetic activity and autonomic dysfunction (Low 1993) and has been repeatedly implicated in the pathogenesis of cardiovascular disease. Increased sympathetic nervous system activity is also associated with depressive disorder in both cardiac and non-cardiac patients (Goldstein and Niaura 1992). A model in which sympathetic tone unifies the mechanisms underlying the relationship between major depression and cardiovascular disease is consistent with the available evidence, as increased sympathetic tone has been reported in both depression (Goldstein and Niaura 1992) and CVD (Barron and Lesh 1996, Hockman et al. 1966, Schwartz 1990, Willich et al. 1993), and because the mechanisms proposed to account for elevated cardiovascular risk in depression appear to be exacerbated by sympathetic nervous system activity. Sympathetic tone has been shown to increase as a result of injections of the AA-derived eicosanoid PGE_2 (Gullner 1983, MacNeil et al. 1997). Circulating NE levels also increase in a dose-dependent manner in response to prostaglandins (Yeragani et al. 1991), in particular the AA-derived PGE_2 (Yokotani et al. 1995).

4.3.1. A model of CNS mechanisms regulating cardiovascular responses to emotional stress

Emotional stressors, including anger, clearly alter autonomic nervous system activity. Preliminary findings indicate that the frontal cortex, limbic circuits, subthalamic and brainstem regions regulate sympathetic and parasympathetic balance during emotional stress, possibly via serotonergic (Lehnert et al. 1987), dopaminergic and noradrenergic pathways (Verrier 1987). The frontal lobe is clearly important in regulating autonomic responses to emotional events. Raine et al. (2000) reported decreased prefrontal gray matter volume and reduced autonomic activity among subjects with antisocial personality disorder. The medial prefrontal cortex appears to have a specific role in linking emotional reactivity to cardiovascular responses; subjects with damage to the medial prefrontal cortex lose the capacity to have autonomic nervous system responses to emotional stimuli (Bechara et al. 1996). The medial prefrontal cortex has significant overlapping serotonergic, dopaminergic and noradrenergic innervation (Goldman-Rakic et al. 1990). One of the final common pathways of autonomic processing in the CNS may be the nucleus tractus solitarius (located in the brainstem) (Callera et al. 1997). Activation (by ketamine) and blockage (by ondansetron) of serotonin (5-HT$_3$) receptors, presumably in the nucleus tractus solitarius, indicate that serotonergic activity is important to the regulation of a sensitive measure of autonomic function, heart-rate variability (DePetrillo et al. 2000). Although the brain stem is a critical region, global changes in serotonergic function may be very important since so many different regions are involved. Rabinowitz and Lown (1978) elevated serotonin concentrations in CNS globally with tryptophan loading, which caused diminished sympathetic neuronal activity and increased the threshold of cardiac ventricular instability by 50%. Lehnert et al. (1987) found similar results with 5-L-hydroxytryptophan loading; a 42% increase in the ventricular fibrillation threshold and suppression of efferent sympathetic activity. We predict that increasing dietary intake and tissue concentrations of DHA and or EPA may cause changes in serotonergic and dopaminergic function in several brain regions simultaneously, including the frontal cortex and brain stem, and will reduce cardiovascular mortality in response to emotional stress.

4.3.2. Emotionally induced increases in sympathetic tone, plasma NE and CVD risk: Exacerbation by EPA/DHA insufficiency

Heightened emotional states are thought to contribute to the development of cardiovascular disease and increase the risk of adverse cardiac events through increases in sympathetic tone and excessive NE release (Booth-Kewley and Friedman 1987, Carney et al. 1993, Ford et al. 1998, Musselman et al. 1998, Sloan et al. 1999, Yeragani et al. 1995). Heightened emotional states such as depression (Fielding 1991), anger (Verrier and Mittleman 1996), and social stress (Kaplan et al. 1991b) are also associated with an increased risk of cardiovascular death, due in part to increased sympathetic activity in relation to parasympathetic activity. The success of psychosocial interventions in reducing cardiovascular mortality (Williams and Littman 1996) further emphasizes the strength of this relationship. These adverse effects of stress may be alleviated by an elevated omega-3 status. In a five-year prospective trial, fish consumption reduced hostility scores (Weidner et al. 1992), and DHA has been shown to reduce hostile responses of students under stress (Sawazaki et al. 1999). In both human and animal studies, DHA and EPA supplementation appears to prevent stress-induced increases in plasma NE (Rousseau et al. 1998, Sawazaki et al. 1999) as well as reduce baseline NE in placebo-controlled trials (Hamazaki et al. 1999). Among healthy volunteers under considerable stress, daily treatment with 1.5 grams of DHA for nine weeks reduced resting plasma NE concentration by 31% (Sawazaki et al. 1999). The plasma epinephrine to norepinephrine ratio increased in every DHA-treated subject (+78%, $p < 0.02$).

In addition to diminishing sympathetic activity and NE release, EPA and DHA protect vascular endothelial tissues and circulating cells from the damage and adverse consequences induced by excessive plasma NE release. NE-induced hemodynamic changes including increased heart rate, vasoconstriction and hypertension can clearly be reduced by increased EPA and DHA tissue (Grimsgaard et al. 1998, Russo et al. 1995). Several large clinical trials have established that fish-oil treatments lower blood pressure in hypertensive patients (Bonaa et al. 1990, Knapp and FitzGerald 1989, Mori et al. 1999). EPA inhibits NE-induced contractions in vascular smooth muscle, which may contribute to its vasorelaxant effect (Engler et al. 1999). Smooth muscle cell proliferation and intimal hyperplasia, which are normally stimulated after endothelial damage, were virtually abolished by pretreatment with EPA and DHA (Pakala et al. 1999c). Thus, variability in omega-3 status may in part explain why subjects undergoing the same amount of stress have different cardiovascular responses.

4.4. Other possible cardiovascular mechanisms

Several additional mechanisms deserve brief mention here although evidence is lacking to clearly establish their relationships to either CVD or depression or a physiological effect of omega-3 fatty acids. These mechanisms include effects of omega-3 essential fatty acids on elevated triglyceride levels, elevated homocysteine levels, and depressed nitric oxide production. Reviews of more that 40 clinical intervention trials describe a robust and uniform effect of fish-oil supplementation in reducing plasma triglyceride concentrations (Harris 1996, 1997). Some investigators have found elevated triglyceride concentrations among depressed patients (Glueck et al. 1993, Kuczmierczyk et al. 1996, Melamed et al. 1997, Morgan et al. 1993) while others have not (Freedman et al. 1995, Olusi and Fido 1996). Elevated triglyceride concentrations have also been suggested as an independent risk factor for cardiovascular morbidity (Roche and Gibney 2000), but this finding is not well established. Thus, data potentially linking elevated triglycerides to both depression and increased cardiovascular risk are still evolving. Abundant data exist linking elevated plasma homocysteine concentrations to cardiovascular disease (Nehler et al. 1997, Nygard et al. 1997). In particular, there appears to exist a graded relationship between homocysteine concentrations and cardiovascular mortality (Refsum et al. 1998), even among patients with pre-established coronary artery disease ($p = 0.01$) (Nygard et al. 1997). However, homocysteine was elevated in the brain tissue of patients with intractable depression in only one report (Francis et al. 1989). Folate- and vitamin-B12-deficiency-induced alterations in homocysteine methylation have been implicated in the pathophysiology of depression (Bottiglieri 1996, Fava et al. 1997). In one study, treatment with 12 grams of fish oil per day for 3 weeks successfully diminished serum homocysteine ($p < 0.01$, Olszewski and McCully 1993). A reduction in nitric oxide synthase activity has also been proposed to link CVD and depression (Finkel et al. 1996), but direct clinical data for this association are sparse. Nitric oxide production is clearly involved in atherosclerotic plaque formation, platelet aggregation and vascular autonomic responses (Connor and Connor 1997), and one study reported lower serum levels of nitric oxide among depressed patients with ischemic heart disease (Finkel et al. 1996). A reduction in the number of nitric-oxide-synthase-containing neurons in the paraventricular nucleus of the hypothalamus of depressed patients has also been reported (Bernstein et al. 1998). Thus, it has been proposed that nitric oxide synthesis may be involved in the cytokine- and immune-induced hyperactivity of the HPA axis in depression (van Amsterdam and Opperhuizen 1999). Consistent with our presumption of omega-3 insufficiency on depression, EPA and DHA appear to strongly enhance nitric oxide production (Harris et al. 1997) and potentially reduce HPA-axis hyperactivity. The data linking elevated triglycerides, elevated homocysteine and reduced nitric oxide levels to CVD and depression are not well substantiated. Nonetheless, these mechanisms do appear to be consistent with an insufficiency of omega-3 fats.

EPA and DHA also inhibit phosphoinositide turnover in peripheral tissues, thus inhibiting another step in thrombus formation. When rabbits were fed diets rich in fish oils, phosphoinositide turnover was significantly reduced when their platelets were exposed to collagen or thrombin (Medini et al. 1990). Production of inositol phosphates (IP, IP_2 and IP_3) stimulated by a TXA_2 analogue was markedly diminished when platelets (Chetty et al. 1989) or neutrophils (Sperling et al. 1993) were preincubated with EPA. EPA pretreatment of cardiac myocytes decreased phenylephrine-induced inositol phosphate production by 33.2% (de Jonge et al. 1996). In vascular smooth muscle cells, pretreatment with fish oil or EPA completely abolished the inositol trisphosphate production stimulated by low-density lipoproteins or by TXA_2 (Locher et al. 1989). Since lithium appears to modulate IP_3 turnover at physiological doses (Berridge 1989), it is tempting to speculate that a mechanism of action of DHA or EPA in affective disorders could be modulation of IP_3 turnover (Hibbeln and Salem 1995, Stoll et al. 1999a). These findings may not extend to the central nervous system. No changes in basal phosphoinositide levels were found in the hippocampi of rats after 3 generations of omega-3 deficiency (Nanjo et al. 1999). Furthermore, concentrations of EPA in neuronal tissue are difficult to detect and, in contrast to peripheral tissues, phosphoinositide phospholipids from neuronal tissues contain little DHA (Ikeda et al. 1986).

Depressed patients may be particularly susceptible to serotonin-mediated platelet activation due to an increased binding density of serotonin type-2 (5-HT$_2$) receptors on platelets (Ikeda et al. 1986). Autopsy studies of the frontal cortex of suicide victims have reported a 44% increase in 5-HT$_2$ receptor number without changes in affinity (Stanley et al. 1983). Strikingly similar findings have been reported in rat frontal cortex in response to omega-3 fat deprivation, which again resulted in a 44% increase in 5-HT$_2$ receptor number without changes in affinity (DeLion et al. 1996). A possible mechanism of the effects of EPA and DHA on 5-HT$_2$ receptor number may be through effects on 5-HT$_2$ mRNA production. In endothelial cells, pre-incubation with EPA and DHA abolished serotonin-induced increases in 5-HT$_2$ mRNA levels (Pakala et al. 1999a). These findings are consistent with the observation that low plasma concentrations of DHA predict low concentrations of cerebrospinal fluid 5-hydroxyindoleacetic acid (CSF 5-HIAA) among healthy controls (Hibbeln et al. 1998a). However, to our knowledge, no studies have examined the effects of EPA and DHA on 5-HT$_2$ binding and density in platelets.

Increased risk for the progression of atherosclerosis is not well documented in depression, and the possible roles for EPA and DHA in the prevention of atherosclerosis are still being explored. Macrophage recruitment into the vascular wall following uptake of oxidized LDL initiates the formation of foam cells and fatty streaks. This is one of the earliest events in atherogenesis (de Winther et al. 2000). This invasion into the subendothelial space depends in part on the endothelial expression of vascular cell adhesion molecule 1, which is induced by IL-1, TNF, and others (De Caterina et al. 1994, 2000). Consistent with these findings, IL-1β and TNF have been shown to induce arteriosclerosis-like changes in the porcine coronary artery (Fukumoto et al. 1997a,b). The expression of endothelial leukocyte adhesion molecule 1, the secretion of inflammatory mediators, and leukocyte adhesion to endothelial cells in response to cytokines are markedly reduced by prior incorporation of DHA into cultured endothelial cells (De Caterina et al. 1994). Thus, an anti-atherosclerotic property of DHA may be diminishing endothelial responses to cytokines, in addition to reduction of cytokine production by EPA and DHA.

4.5. Summary: CVD, depression and low omega-3 status

An insufficiency of omega-3 fatty acids, specifically EPA and DHA, aggravates every risk factor that has been described as linking major depression to an increased risk of cardiovascular mortality. Specifically, major depression has been associated with increased platelet reactivity, elevated plasma NE concentrations, hypertension, low heart-rate variability and HPA-axis and sympathetic hyperactivity. The strongest evidence of increased cardiovascular mortality among depressed patients is the increased risk of sudden cardiac death and vulnerability to arrhythmias. The cellular mechanisms through which EPA and DHA reduce risk of cardiac arrhythmias have been elucidated, and their ability to reduce risk of sudden cardiac death has been documented in both clinical intervention trials and ecological studies. Both are aggravated by low tissue concentrations of EPA and DHA. There exist substantial data that increasing EPA and DHA status may abate each of these processes. The proposition that omega-3 insufficiency is a causal link between depression and cardiovascular mortality should be examined in clinical and epidemiological trials.

5. Low cholesterol, cholesterol lowering and increased risk of mortality or depression

We posit that the nutritional status of omega-3 essential fatty acids may be an important determinant factor influencing the relationship between low cholesterol status and increased risk of suicidal, homicidal or depressive behaviors (Hibbeln and Salem 1995). Several investigators have reported that low serum cholesterol concentrations are associated with a greater risk of mortality in observational studies (see other chapters). However, this association does not necessarily demonstrate that lowering serum cholesterol will increase depression or suicidal behavior. In fact, cholesterol reduction was not linked to non-illness mortality in a meta-analysis of randomized clinical trials (Muldoon et al. 2001). An important consideration in examining the psychotropic effects of cholesterol lowering is to consider the secondary effects on essential fatty acid status. For example, a diet that is lower in fat, but contains large amounts of safflower or corn oil, can markedly lower EPA and DHA concentrations (Hibbeln and Salem 1995, Kaplan et al. 1991a). In addition to low-fat diets, cholesterol-lowering drugs themselves can have secondary effects of lower polyunsaturate status, as is the case with fibrates, or increasing polyunsaturate status as is the case with statins (Hibbeln and Salem 1996).

One hypothesis of the mechanism linking increased risk of non-illness mortality to cholesterol lowering has been frequently referenced (Endelberg 1992). Endelberg postulated that low serum cholesterol concentrations are a marker of low brain cholesterol concentrations. He predicted that lowering serum cholesterol ultimately will alter the biophysical properties of synaptic membranes, change the shape of serotonergic receptors and alter serotonergic function. Unfortunately, the central flaw of this hypothesis is that the brain makes its own cholesterol de novo, and cholesterol in plasma or serum does not cross the blood–brain barrier (Edmond et al. 1991, Pardridge and Mietus 1980). In contrast, omega-3 polyunsaturated fatty acids have an important role in determining membrane biophysical properties, and the brain is dependent upon dietary sources of the essential fatty acids. Depletions of highly unsaturated fatty acids from neuronal membranes may have more important biophysical consequences in modulating serotonergic neurotransmission than alterations in cholesterol (Hibbeln et al. 2000d). These issues will be discussed in detail below.

The biophysical hypothesis put forth by Endelberg (Endelberg 1992) seems biologically implausible for total cholesterol, but not for highly unsaturated essential fatty acids, in light of the following observations:

(1) Under normal human dietary conditions, altering dietary cholesterol intake has only a marginal effect on plasma cholesterol concentrations. Intake of saturated, mono-unsaturated and polyunsaturated fatty acids have a much greater role in determining plasma cholesterol concentrations in large epidemiological studies of human populations (Keys 1997).

(2) Brain cholesterol concentrations are largely unrelated to plasma cholesterol concentrations. A few early studies in the 1950s and 1960s suggested that cholesterol might be able to cross the blood–brain barrier (Dobbing 1963) but recent well-controlled studies have demonstrated that these results were probably artifactual (Pardridge and Mietus 1980). Cholesterol could not be detected crossing into the brain despite controlling for preparation artifacts and equilibration with endogenous serum preparations (Pardridge and Mietus 1980). Artificial milk which contained either extremely high, low or normal concentrations of dietary cholesterol profoundly altered plasma and lung cholesterol concentrations, but brain concentrations of cholesterol and sterols were unaltered (Edmond et al. 1991). During development, when the most rapid accumulation of cholesterol in the brain occurs, the brain is able to synthesize in situ all the cholesterol required (Edmond et al. 1991, Jones et al. 1975a,b). Thus, alterations in dietary intake of cholesterol or in plasma cholesterol concentrations are not likely to influence biophysical properties of neuronal membranes.

In contrast to cholesterol, the brain is entirely dependent upon dietary or maternal sources for polyunsaturated essential fatty acids (Salem 1989), and selectively concentrates docosahexaenoic acid into synaptic neuronal membranes (Connor et al. 1990). Dietary deficiencies of n-3 essential fatty acids during development result in deficiencies of docosahexaenoic acid in brain tissues (Farquharson et al. 1992).

(3) Although cholesterol is important in determining membrane order parameters, polyunsaturate composition is a powerful modulator of cholesterol-induced changes in membrane order. Cholesterol-induced membrane condensation, measured by increasing order in hydrocarbon chains, is reduced with greater unsaturation of surrounding acyl chains. Membranes containing docosahexaenoic acid, a highly unsaturated fatty acid, showed the lowest amount of cholesterol-induced condensation (Barry and Gawrisch 1995). One study in particular that examined the effects of cholesterol on serotonergic receptors is often referenced (Heron et al. 1980). However, in addition to cholesterol, linoleic acid, a polyunsaturated essential fatty acid with two double bonds, reduced membrane viscosity and decreased the number of high-affinity sites on serotonin receptors to a greater degree than cholesterol (Heron et al. 1980). More detailed biophysical studies of this superfamily of G protein-coupled receptors, e.g. rhodopsin, which share both structural and functional properties, have been conducted. Mitchell et al. (1992b) demonstrated that the membrane phospholipid acyl chain composition was the primary determinant of the relationship between bulk membrane packing properties and the formation of metarhodopsin II (meta II), the active form of the receptor. Cholesterol was found to have a secondary, modulating effect. In addition, highly unsaturated phospholipids such as those containing di-docosahexaenoate promoted the formation of meta II and markedly diminished cholesterol-induced inhibition of meta II formation. Highly unsaturated fatty acids, especially docosahexaenoic acid, contribute unique biophysical properties to neuronal membranes such as lateral compressibility and lateral domain formation which influence receptor conformation and signal transduction (Litman et al. 1991, Mitchell et al. 1992a,b, 1996). Mitchell and Litman (1998) used time-resolved fluorescence emission and decay of fluorescence

anisotropy of 1,6-diphenyl-1,3,5-hexatriene (DPH) to characterize equilibrium and dynamic structural properties of bilayers containing 30 mol% cholesterol and either saturated, mono-unsaturated or polyunsaturated fatty acids in phospholipids. The lifetime distributions were dramatically narrowed by the addition of cholesterol in all bilayers except the two consisting of dipolyunsaturated phospholipids. The effect of cholesterol was especially diminished in di-22:6 n-3 phosphotidylcholine, suggesting that this phospholipid may be particularly effective at promoting lateral domains, which are cholesterol-rich and unsaturation-rich, respectively.

(4) Animal studies have shown conflicting results when examining the effects of dietary changes in cholesterol intake on brain serotonin concentrations. When cholesterol intake was lowered as an isolated dietary variable, rhesus monkeys exhibited more violent behavior, had significantly lower levels of CSF 5-HIAA and had lower serum cholesterol (235 mg/dl), on the low-cholesterol diet compared to a cholesterol-rich diet (623 mg/dl) (Kaplan et al. 1991a). Kaplan also reported that low-fat diets (with an decreased omega-3 composition) lowered CSF 5-HIAA concentrations in rhesus monkeys (Kaplan et al. 1994). Consistent with these observations, Malyszko et al. (1994) found that serum cholesterol concentrations positively predicted CSF serotonin concentrations among non-human primates. However, in gerbils, no differences in brain tryptophan, serotonin or 5-HIAA concentrations were found as a function of circulating cholesterol when concentrations were in a range of a 1.5 to 20 mol/ml (Fernstrom et al. 1996). In contrast, rats maintained on an n-3 deficient diet, which reduced concentrations of docosahexaenoic acid in brain tissues, showed a 44% increase in 5-HT$_2$ receptor number in the frontal cortex (DeLion et al. 1996).

(5) In our review of the literature, we have found mostly negative reports of associations between measures of cholesterol and central serotonergic function in human studies. One important caveat is that peripheral measures of serotonin are poor predictors of central serotonergic function (Mann et al. 1992). Contrary to the prediction of Endelberg (1992), Engstrom et al. (1995) found that high-density lipoprotein cholesterol showed positive correlations with the dopamine metabolite homovanillic acid ($r = 0.39$, $p = 0.04$) and the serotonin metabolite 5-HIAA ($r = 0.34$, $p = 0.07$). However, the suicide intent scales and hopelessness scales did not correlate significantly with serum lipids. In comparing 20 healthy controls to 20 subjects on cholesterol-lowering therapies no differences were found in prolactin response to D-fenfluramine, serotonin content of platelets or plasma serotonin levels (Delva et al. 1996). Comparing suicide attempters and non-attempters (Almeida-Montes et al. 2000) no differences in plasma total cholesterol, LDL and triglycerides were reported, but lower plasma serotonin and tryptophan concentrations were found among the suicide attempters. Cholesterol measures did not predict plasma serotonergic measures. Buydens-Branchey et al. (2000) found significantly lower levels of HDL cholesterol in patients who had a history of aggression. Lower levels of HDL cholesterol were also found to be significantly associated with more intense 'high' and 'activation-euphoria' responses as well as with blunted cortisol responses to m-CPP. Terao et al. (1997, 2000) reported positive correlations between serum cholesterol levels and hormonal responses to meta-chlorophenylpiperazine administration. These results suggest that serum cholesterol levels may be positively associated with

serotonergic receptor function. In summary, while peripheral measures of cholesterol poorly predict CSF 5-HIAA measures, there may be some relationship between plasma cholesterol measures and responses to serotinergic agonists. Since these responses are receptor mediated, the role of the subjects' omega-3 fatty acid status should also be considered.

As reviewed elsewhere in this book, genetic variants in apolipoproteins or their receptors could account for lower plasma cholesterol concentrations in groups of violent or impulsive subjects. Corrigan et al. (1997) reported that violent prisoners have higher concentrations of apo A-IV ($p < 0.000001$) and apo E ($p < 0.0002$). These apolipoproteins are involved in the transport of both cholesterol and polyunsaturated fatty acids. In high-density lipoproteins, omega-3 fatty acids predicted Apo E concentrations in the offender group, but not in the control group. Corrigan et al. (1997) also reported low concentrations of plasma docosahexaenoic acid from these violent offenders, which replicated the findings of Virkkunen et al. (1987) who reported low concentrations of docosahexaenoic acid in violent offenders with antisocial personality disorder. These findings may suggest that abnormalities in plasma cholesterol concentrations and polyunsaturated fatty acid transportation may be linked by abnormalities in apolipoprotein metabolism. A continued examination into allelic differences in subgroups may eventually explain apparently contradictory data such as the report of lower cholesterol among suicide attempters (Garland et al. 2000) in contrast to reports of higher total cholesterol concentrations among violent suicide attempters (Tanskanen et al. 2000). We stress that the role of polyunsaturated fatty acids and measure of cholesterol should not be examined in isolation in future studies of the roles these lipids may play in depression violence or cardiovascular disease.

References

Abeywardena, M.Y., McLennan, P.L. and Charnock, J.S. (1992) Role of eicosanoids in dietary fat modification of cardiac arrhythmia and ventricular fibrillation. In: A. Sinclair and R. Gibson (Eds.), Essential Fatty Acids and Eicosanoids: Invited Papers from the Third International Congress. American Oil Chemists' Society, Champaign, IL, pp. 257–261.

Adams, P.B., Lawson, S., Sanigorski, A. and Sinclair, A.J. (1996) Arachidonic acid to eicosapentaenoic acid ratio in blood correlates positively with clinical symptoms of depression. Lipids 31, S157–S161.

Agostoni, C., Riva, E., Trojan, S., Bellu, R. and Giovannini, M. (1995) Docosahexaenoic acid status and developmental quotient of healthy term infants. Lancet 346, 638.

Ahern, D.K., Gorkin, L., Anderson, J.L., Tierney, C., Hallstrom, A., Ewart, C., Capone, R.J., Schron, E., Kornfeld, D. and Herd, J.A. (1990) Biobehavioral variables and mortality or cardiac arrest in the Cardiac Arrhythmia Pilot Study (CAPS). Am. J. Cardiol. 66, 59–62.

Akselrod, S., Gordon, D., Madwed, J.B., Snidman, N.C., Shannon, D.C. and Cohen, R.J. (1985) Hemodynamic regulation: investigation by spectral analysis. Am. J. Physiol. 249, H867–H875.

Al, M., Houwelingen, A.C. and Honstra, G. (1994) Docosahexaenoic acid, 22:6(n3), cervonic acid (CA), and hypertension in pregnancy: consequences for mother and child. In: Abstracts, Scientific Conf. on Omega-3 Fatty Acids in Nutrition, Vascular Biology and Medicine, American Heart Association, Houston, TX, Abstract 56, pp. 17–19.

Albert, C.M., Hennekens, C.H., O'Donnell, C.J., Ajani, U.A., Carey, V.J., Willett, W.C., Ruskin, J.N. and Manson, J.E. (1998) Fish consumption and risk of sudden cardiac death. J. Am. Med. Assoc. 279, 23–28.

Almeida-Montes, L.G., Valles-Sanchez, V., Moreno-Aguilar, J., Chavez-Balderas, R.A., Garcia-Marin, J.A., Sotres, J.F. and Heinze-Martin, G. (2000) Relation of serum cholesterol, lipid, serotonin and tryptophan levels to severity of depression and to suicide attempts. J. Psychiatry Neurosci. 25, 371–377.

American Psychiatric Association (1994) Diagnostic Criteria from DSM-IV, American Psychiatric Association Press, Washington, DC.

Anfossi, G. and Trovati, M. (1996) Role of catecholamines in platelet function: pathophysiological and clinical significance. Eur. J. Clin. Invest. 26, 353–370.

Apter, A., Plutchik, R. and van Praag, H.M. (1993) Anxiety, impulsivity and depressed mood in relation to suicidal and violent behavior. Acta Psychiatr. Scand. 87, 1–5.

Ascherio, A., Rimm, E.B., Giovannucci, E.L., Spiegelman, D., Stampfer, M. and Willett, W.C. (1996) Dietary fat and risk of coronary heart disease in men: cohort follow up study in the United States. Br. Med. J. 313, 84–90.

Austead, N., Innis, S.M. and de la Presa Owens, S. (2000) Auditory evoked response and brain phospholipids fatty acids and monoamines in rats fed formula with or without arachidonic acid (AA) and/or docosahexaenoic acid (DHA) (abstract). In: Brain Uptake and Utilization of Fatty Acids. Applications to peroxisomal biogenesis disorders. National Institutes of Health, Bethesda, MD, March 2–4, 2000.

Bailey, J.R., Fitzgerald, D.M. and Applegate, R.J. (1996) Effects of constant cardiac autonomic nerve stimulation on heart rate. Am. J. Physiol. 270, H2081–H2087.

Balogh, S., Fitzpatrick, D.F., Hendricks, S.E. and Paige, S.R. (1993) Increases in heart rate variability with successful treatment in patients with major depressive disorder. Psychopharmacol. Bull. 29, 201–206.

Banki, C.M., Karmacsi, L., Bissette, G. and Nemeroff, C.B. (1992) CSF corticotropin-releasing hormone and somatostatin in major depression: response to antidepressant treatment and relapse. Eur. Neuropsychopharmacol. 2, 107–113.

Barron, H.V. and Lesh, M.D. (1996) Autonomic nervous system and sudden cardiac death. J. Am. Coll. Cardiol. 27, 1053–1060.

Barry, J.A. and Gawrisch, K. (1995) Effects of ethanol on lipid bilayers containing cholesterol, gangliosides, and sphingomyelin. Biochemistry 34, 8852–8860.

Bayes de Luna, A., Coumel, P. and Leclercq, J.F. (1989) Ambulatory sudden cardiac death: mechanisms of production of fatal arrhythmia on the basis of data from 157 cases. Am. Heart J. 117, 151–159.

Bebbington, P. and Ramana, R. (1995) The epidemiology of bipolar affective disorder. Soc. Psychiatry Psychiatr. Epidemiol. 30, 279–292.

Bechara, A., Tranel, D., Damasio, H. and Damasio, A.R. (1996) Failure to respond autonomically to anticipated future outcomes following damage to prefrontal cortex. Cereb. Cortex 6, 215–225.

Benedict, C.R., Shelton, B., Johnstone, D.E., Francis, G., Greenberg, B., Konstam, M., Probstfield, J.L. and Yusuf, S. (1996) Prognostic significance of plasma norepinephrine in patients with asymptomatic left ventricular dysfunction. SOLVD Investigators. Circulation 94, 90–697.

Bernstein, H.G., Stanarius, A., Baumann, B., Henning, H., Krell, D., Danos, P., Falkai, P. and Bogerts, B. (1998) Nitric oxide synthase-containing neurons in the human hypothalamus: reduced number of immunoreactive cells in the paraventricular nucleus of depressive patients and schizophrenics. Neuroscience 83, 867–875.

Berridge, M.J. (1989) The Albert Lasker Medical Awards. Inositol trisphosphate, calcium, lithium, and cell signaling. J. Am. Med. Assoc. 262, 1834–1841.

Bible King James Version, Oxford Text Archive.

Bigger, J.T., Fleiss, J.L., Rolnitzky, L.M. and Steinman, R.C. (1993) The ability of several short-term measures of RR variability to predict mortality after myocardial infarction. Circulation 88, 927–934.

Bigger Jr, J.T., Kleiger, R.E., Fleiss, J.L., Rolnitzky, L.M., Steinman, R.C. and Miller, J.P. (1988) Components of heart rate variability measured during healing of acute myocardial infarction. Am. J. Cardiol. 61, 208–215.

Billman, G.E., Schwartz, P.J. and Stone, H.L. (1982) Baroreceptor reflex control of heart rate: a predictor of sudden cardiac death. Circulation 66, 874–880.

Birch, E.E., Garfield, S., Hoffman, D.R., Uauy, R. and Birch, D.G. (2000) A randomized controlled trial of early dietary supply of long-chain polyunsaturated fatty acids and mental development in term infants. Dev. Med. Child Neurol. 42, 174–181.

Bjorneboe, A. and Bjorneboe, G.E. (1993) Antioxidant status and alcohol-related diseases. Alcohol Alcoholism 28, 111–116.

Block, E. and Edwards, D. (1987) Effects of plasma membrane fluidity on serotonin transport by endothelial cells. Am. J. Physiol. 253, C672–C678.

Bonaa, K.H., Bjerve, K.S., Straume, B., Gram, I.T. and Thelle, D. (1990) Effect of eicosapentaenoic and docosahexaenoic acids on blood pressure in hypertension. A population-based intervention trial from the Tromso study. N. Engl. J. Med. 322, 795–801.

Booth-Kewley, S. and Friedman, H.S. (1987) Psychological predictors of heart disease: A quantitative review. Psychol. Bull. 101, 343–362.

Bottiglieri, T. (1996) Folate, vitamin B12, and neuropsychiatric disorders. Nutr. Rev. 54, 382–390.

Bourre, J.M., Bonneil, M., Dumont, O., Piciotti, M., Nalbone, G. and Lafont, H. (1988) High dietary fish oil alters the brain polyunsaturated fatty acid composition. Biochim. Biophys. Acta 960, 458–461.

Broadhurst, C., Cunnane, S. and Crawford, M. (1998) Rift valley lake fish and shellfish provided brain-specific nutrition for early Homo. Br. J. Nutr. 79, 3–21.

Brown, A.S., van Os, J., Driessens, C., Hoek, H.W. and Susser, E.S. (2000) Further evidence of relation between prenatal famine and major affective disorder. Am. J. Psychiatry 157, 190–195.

Bruhn, J.G., Paredes, A., Adsett, C.A. and Wolf, S. (1974) Psychological predictors of sudden death in myocardial infarction. J. Psychosom. Res. 18, 187–191.

Bugajski, J. (1996) Role of prostaglandins in the stimulation of the hypothalamic–pituitary–adrenal axis by adrenergic and neurohormone systems. J. Physiol. Pharmacol. 47, 559–575.

Burgess, J.R. (1998) Attention deficit hyperactivity disorder, observational and interventional studies. In: NIH Workshop on Omega-3 Essential Fatty Acids and Psychiatric Disorders, National Institutes of Health, Bethesda, MD, p. 23. Abstract.

Burton, R. (1988) The Anatomy of Melancholy, The Classics of Psychiatry and Behavioral Sciences Library, Division of Gryphon Editions, Birmingham, AL.

Buydens-Branchey, L., Branchey, M., Hudson, J. and Fergeson, P. (2000) Low HDL cholesterol, aggression and altered central serotonergic activity. Psychiatry Res. 93, 93–102.

Callera, J.C., Sevoz, C., Laguzzi, R. and Machado, B.H. (1997) Microinjection of a serotonin3 receptor agonist into the NTS of unanesthetized rats inhibits the bradycardia evoked by activation of the baro- and chemoreflexes. J. Auton. Nerv. Syst. 63, 127–136.

Campbell, F.M., Gordon, M.J. and Dutta-Roy, A.K. (1998) Placental membrane fatty acid-binding protein preferentially binds arachidonic and docosahexaenoic acids. Life Sci. 63, 235–240.

Carlson, S.E., Cooke, R.J., Rhodes, P.G., Peeples, J.M., Werkman, S.H. and Tolley, E.A. (1991) Long-term feeding of formulas high in linolenic acid and marine oil to very low birth weight infants: phospholipid fatty acids. Pediatr. Res. 30, 404–412.

Carney, R.M., Rich, M.W., teVelde, A., Saini, J., Clark, K. and Jaffe, A.S. (1987) Major depressive disorder in coronary artery disease. Am. J. Cardiol. 60, 1273–1275.

Carney, R.M., Freedland, K.E., Rich, M.W., Smith, L.J. and Jaffe, A.S. (1993) Ventricular tachycardia and psychiatric depression in patients with coronary artery disease. Am. J. Med. 95, 23–28.

Carney, R.M., Saunders, R.D., Freedland, K.E., Stein, P., Rich, M.W. and Jaffe, A.S. (1995) Association of depression with reduced heart rate variability in coronary artery disease. Am. J. Cardiol. 76, 562–564.

Carney, R.M., Freedland, K.E., Veith, R.C., Cryer, P.E., Skala, J.A., Lynch, T. and Jaffe, A.S. (1999) Major depression, heart rate, and plasma norepinephrine in patients. Biol. Psychiatry 45, 458–463.

Casacchia, M., Meco, G., Pirro, R., Di Cesare, E., Allegro, A., Cusimano, G. and Marola, W. (1982) Phospholipid liposomes in depression: a double blind study versus placebo. Int. Pharmacopsychiatry 17, 274–279.

Cenacchi, T., Bertoldin, T., Faarina, C., Fiori, M.G. and Crepaldi, G. (1993) Cognitive decline in the elderly: a double-blind placebo-controlled multicenter study on efficacy of phosphatidyl-serine administration. Aging Clin. Exp. Res. 5, 123–133.

Chalon, S., Delion-Vancassel, S., Belzung, C., Guilloteau, D., Leguisquet, A.M., Besnard, J.C. and Durand, G. (1998) Dietary fish oil affects monoaminergic neurotransmission and behavior in rats. J. Nutr. 128, 2512–2519.

Champoux, M., Hibbeln, J., Higley, J.D., Shannon, C., Purple, M., Salem Jr, N. and Suomi, S.J. (2001) Formula supplementation with DHA and AA and improvements in neonatal development in rhesus monkeys. Pediatr. Res., in press.

Chetty, N., Vickers, J.D., Kinlough-Rathbone, R.L., Packham, M.A. and Mustard, J.F. (1989) Eicosapentaenoic acid interferes with U46619-stimulated formation of inositol phosphates in washed rabbit platelets. Thromb. Haemost. 62, 1116–1120.

Christensen, J.H., Gustenhoff, P., Korup, E., Aaroe, J., Toft, E., Moller, J., Rasmussen, K., Dyerberg, J. and Schmidt, E.B. (1996) Effect of fish oil on heart rate variability in survivors of myocardial infarction: a double blind randomized controlled trial. Br. Med. J. 312, 677–678.

Christensen, J.H., Korup, E., Aaroe, J., Toft, E., Moller, J., Rasmussen, K., Dyerberg, J. and Schmidt, E.B. (1997) Fish consumption, n-3 fatty acids in cell membranes, and heart rate variability in survivors of myocardial infarction with left ventricular dysfunction. Am. J. Cardiol. 79, 1670–1673.

Christensen, J.H., Aaroe, J., Knudsen, N., Dideriksen, K., Kornerup, H.J., Dyerberg, J. and Schmidt, E.B. (1998) Heart rate variability and n-3 fatty acids in patients with chronic renal failure – a pilot study. Clin. Nephrol. 49, 102–106.

Christensen, J.H., Christensen, M.S., Dyerberg, J. and Schmidt, E.B. (1999) Heart rate variability and fatty acid content of blood cell membranes: a dose–response study with n-3 fatty acids. Am. J. Clin. Nutr. 70, 331–337.

Connor, S.L. and Connor, W.E. (1997) Are fish oils beneficial in the prevention and treatment of coronary artery disease? Am. J. Clin. Nutr. 66, 1020S–1031S.

Connor, W.E., Neuringer, M. and Lin, D.S. (1990) Dietary effects on brain fatty acid composition: the reversibility of n-3 fatty acid deficiency and turnover of docosahexaenoic acid in the brain, erythrocytes and plasma of rhesus monkeys. J. Lipid Res. 31, 237–247.

Corrigan, F.M., Gray, R.F., Skinner, E.R., Strathdee, A. and Horrobin, D. (1997) Plasma lipoproteins and apolipoproteins in individuals convicted of violent crimes. In: M. Hillbrand and R.T. Spitz (Eds.), Lipids, Health, and Human Behavior, American Psychological Association Press, Washington, DC, pp. 167–178.

Cross-National Collaborative Group (1992) The changing rate of major depression: Cross national comparisons. J. Am. Med. Assoc. 268, 3098–3105.

De Caterina, R., Cybulsky, M.I., Clinton, S.K., Gimbrone Jr, M.A. and Libby, P. (1994) The omega-3 fatty acid docosahexaenoate reduces cytokine-induced expression of proatherogenic and proinflammatory proteins in human endothelial cells. Arterioscler. Thromb. 14, 1829–1836.

De Caterina, R., Liao, J.K. and Libby, P. (2000) Fatty acid modulation of endothelial activation. Am. J. Clin. Nutr. 71, 213S–223S.

De Clerck, F. (1991) Effects of serotonin on platelets and blood vessels. J. Cardiovasc. Pharmacol. 17(Suppl. 5), S1–S5.

de Jonge, H.W., Dekkers, D.H., Bastiaanse, E.M., Bezstarosti, K., van der Laarse, A. and Lamers, J.M. (1996) Eicosapentaenoic acid incorporation in membrane phospholipids modulates receptor-mediated phospholipase C and membrane fluidity in rat ventricular myocytes in culture. J. Mol. Cell. Cardiol. 28, 1097–1108.

de la Presa Owens, S. and Innis, S.M. (1999) Docosahexaenoic and arachidonic acid prevent a decrease in dopaminergic and serotoninergic neurotransmitters in frontal cortex caused by a linoleic and alpha-linolenic acid deficient diet in formula-fed piglets. J. Nutr. 129, 2088–2093.

de la Presa-Owens, S., Innis, S.M. and Rioux, F.M. (1998) Addition of triglycerides with arachidonic acid or docosahexaenoic acid to infant formula has tissue- and lipid class-specific effects on fatty acids and hepatic desaturase activities in formula-fed piglets. J. Nutr. 128, 1376–1384.

de Winther, M.P., van Dijk, K.W., Havekes, L.M. and Hofker, M.H. (2000) Macrophage scavenger receptor class A: A multifunctional receptor in atherosclerosis. Arterioscler. Thromb. Vasc. Biol. 20, 290–297.

DeLion, S., Chalon, S., Guilloteau, D., Besnard, J.C. and Durand, G. (1996) alpha-Linolenic acid dietary deficiency alters age-related changes of dopaminergic and serotoninergic neurotransmission in the rat frontal cortex. J. Neurochem. 66, 1582–1591.

DeLion, S., Chalon, S., Guilloteau, D., Lejeune, B., Besnard, J.C. and Durand, G. (1997) Age-related changes in phospholipid fatty acid composition and monoaminergic neurotransmission in the hippocampus of rats fed a balanced or an n-3 polyunsaturated fatty acid-deficient diet. J. Lipid Res. 38, 680–689.

Delva, N.J., Matthews, D.R. and Cowen, P.J. (1996) Brain serotonin (5-HT) neuroendocrine function in patients taking cholesterol-lowering drugs. Biol. Psychiatry 39, 100–106.

DePetrillo, P.B., Bennett, A.J., Speers, D., Suomi, S.J., Shoaf, S.E., Karimullah, K. and Dee Higley, J. (2000) Ondansetron modulates pharmacodynamic effects of ketamine on electrocardiographic signals in rhesus monkeys. Eur. J. Pharmacol. 391, 113–119.

Deshpande, S. and Akhtar, M. (1997) Sudden cardiac death: magnitude of the problem. In: S. Dunbar, K. Ellenbogen and A. Epstein (Eds.), Sudden Cardiac Death: Past, Present, and Future, Futura Publishing Company, Armonk, NY, pp. 1–27.

Dobbing, J. (1963) The entry of cholesterol into rat brain during development. J. Neurochem. 10, 739–742.

Dolecek, T.A. and Granditis, G. (1991) Dietary polyunsaturated fatty acids and mortality in the Multiple Risk Factor Intervention Trial (MRFIT). World Rev. Nutr. Diet. 66, 205–216.

Drago, F., Continella, G., Mason, G., Hernandez, D. and Scapagnini, U. (1985) Phospholipid liposomes potentiate the inhibitory effect of antidepressant drugs on immobility of rats in a despair test (constrained swim). Eur. J. Pharmacol. 115, 179–184.

Dyerberg, J. and Bang, H.O. (1979) Haemostatic function and platelet polyunsaturated fatty acids in Eskimos. Lancet 2(8140), 433–435.

Eaton, S.B. and Konner, M. (1985) Paleolithic nutrition. A consideration of its nature and current implications. N. Engl. J. Med. 312, 283–289.

Eaton, S.B., Sinclair, A.J., Cordain, L. and Mann, N.J. (1998) Dietary intake of long-chain polyunsaturated fatty acids during the paleolithic. World Rev. Nutr. Diet 83, 12–23.

Edmond, J., Korsak, R.A., Morro, J.W., Torok-Both, G. and Catlin, D.H. (1991) Dietary cholesterol and the origin of cholesterol in the brain of developing rats. J. Nutr. 121, 1323–1330.

Edwards, R., Peet, M., Shay, J. and Horrobin, D. (1998a) Depletion of docosahexaenoic acid in red blood cell membranes of depressive patients. Biochem. Soc. Trans. 26, S142.

Edwards, R., Peet, M., Shay, J. and Horrobin, D. (1998b) Omega-3 polyunsaturated fatty acid levels in the diet and in red blood cell membranes of depressed patients. J. Affect. Disord. 48, 149–155.

Ellis, F.R. and Sanders, T.A.B. (1977) Long chain polyunsaturated fatty acids in endogenous depression. J. Neurol. Neurosurg. Psychiatry 40, 168–169.

Endelberg, H. (1992) Low cholesterol and suicide. Lancet 339, 727–729.

Endres, S., Ghorbani, R., Kelly, V.E., Georgilis, K., Lonnemann, G., van der Meer, J.W.M., Cannon, J.G., Rogers, T.S., Klempner, M.S., Weber, P.C., Schaefer, E.J., Wolff, S.M. and Dinarello, C.A. (1989) The effect of dietary supplementation with n-3 polyunsaturated fatty acids on the synthesis of interleukin 1 and tumor necrosis factor by mononuclear cells. N. Engl. J. Med. 320, 265–271.

Engler, M.B., Ma, Y.H. and Engler, M.M. (1999) Calcium-mediated mechanisms of eicosapentaenoic acid-induced relaxation in hypertensive rat aorta. Am. J. Hypertens. 12, 1225–1235.

Engstrom, G., Alsen, M., Regnell, G. and Traskman-Bendz, L. (1995) Serum lipids in suicide attempters. Suicide Life Threat Behav. 25, 393–400.

Espersen, G.T., Grunnet, N., Lervang, H.H., Nielsen, G.L., Thomsen, B.S., Faarvang, K.L., Dyerberg, J. and Ernst, E. (1992) Decreased interleukin-1 beta levels in plasma from rheumatoid arthritis patients after dietary supplementation with n-3 polyunsaturated fatty acids. Clin. Rheumatol. 11, 393–395.

Farquharson, J., Cockburn, F., Patrick, W.A., Jamieson, E.C. and Logan, R.W. (1992) Infant cerebral cortex phospholipid fatty-acid composition and diet. Lancet 340, 810–813.

Farrell, T.G., Bashir, Y., Cripps, T., Malik, M., Poloniecki, J., Bennett, E.D., Ward, D.E. and Camm, A.J. (1991) Risk stratification for arrhythmic events in postinfarction patients based on heart rate variability, ambulatory electrocardiographic variables and the signal-averaged electrocardiogram. J. Am. Coll. Cardiol. 18, 687–697.

Fava, M., Borus, J.S., Alpert, J.E., Nierenberg, A.A., Rosenbaum, J.F. and Bottiglieri, T. (1997) Folate, vitamin B12, and homocysteine in major depressive disorder. Am. J. Psychiatry 154, 426–428.

Fehily, A.M.A., Bowey, O.A.M., Ellis, F.R., Meade, B.W. and Dickerson, J.W.T. (1981) Plasma and erythrocyte membrane long chain fatty acids in endogenous depression. Neurochem. Int. 3, 37–42.

Fernstrom, M.H., Verrico, C.D., Ebaugh, A.L. and Fernstrom, J.D. (1996) Diet-related changes in serum cholesterol concentrations do not alter tryptophan hydroxylase rate or serotonin concentrations in gerbil brain. Life Sci. 58, 1433–1444.

Fielding, R. (1991) Depression and acute myocardial infarction: a review and reinterpretation. Soc. Sci. Med. 32, 1017–1028.

Finkel, M.S., Laghrissi Thode, F., Pollock, B.G. and Rong, J. (1996) Paroxetine is a novel nitric oxide synthase inhibitor. Psychopharmacol. Bull. 32, 653–658.

Ford, D.E., Mead, L.A., Chang, P.P., Cooper-Patrick, I., Wang, N.Y. and Klag, M.J. (1998) Depression is a risk factor for coronary artery disease in men: the precursors study. Arch. Intern. Med. 158, 1422–1426.

France, R.D., Urban, B., Krishnan, K.R., et al. (1988) CSF corticotropin-releasing factor-like immunoactivity in chronic pain patients with and without major depression. Biol. Psychiatry 23, 86–88.

Francis, P.T., Poynton, A., Lowe, S.L., Najlerahim, A., Bridges, P.K., Bartlett, J.R., Procter, A.W., Bruton, C.J. and Bowen, D.M. (1989) Brain amino acid concentrations and Ca^{2+}-dependent release in intractable depression assessed antemortem. Brain Res. 494, 315–324.

Frasure-Smith, N., Lesperance, F. and Talajic, M. (1995) Depression and 18-month prognosis after myocardial infarction. Circulation 91, 999–1005.

Freedman, D.S., Byers, T., Barrett, D.H., Stroup, N.E., Eaker, E. and Monroe-Blum, H. (1995) Plasma lipid levels and psychologic characteristics in men. Am. J. Epidemiol. 141, 507–517.

Fukumoto, Y., Shimokawa, H., Ito, A., Kadokami, T., Yonemitsu, Y., Aikawa, M., Owada, M.K., Egashira, K., Sueishi, K., Nagai, R., Yazaki, Y. and Takeshita, A. (1997a) Inflammatory cytokines cause coronary arteriosclerosis-like changes and alterations in the smooth-muscle phenotypes in pigs. J. Cardiovasc. Pharmacol. 29, 222–231.

Fukumoto, Y., Shimokawa, H., Kozai, T., Kadokami, T., Kuwata, K., Yonemitsu, Y., Kuga, T., Egashira, K., Sueishi, K. and Takeshita, A. (1997b) Vasculoprotective role of inducible nitric oxide synthase at inflammatory coronary lesions induced by chronic treatment with interleukin-1β in pigs in vivo. Circulation 96, 3104–3111.

Gallai, V., Sarchielli, P., Trequattrini, A., Franceschini, M., Floridi, A., Firenze, C., Alberti, A., Di Benedetto, D. and Stragliotto, E. (1995) Cytokine secretion and eicosanoid production in the peripheral blood mononuclear cells of MS patients undergoing dietary supplementation with n-3 polyunsaturated fatty acids. J. Neuroimmunol. 56, 143–153.

Garland, M., Hickey, D., Corvin, A., Golden, J., Fitzpatrick, P., Cunningham, S. and Walsh, N. (2000) Total serum cholesterol in relation to psychological correlates in parasuicide. Br. J. Psychiatry 177, 77–83.

Gianelli, A., Rabboni, M., Zarattini, F., Malgeri, C. and Magnolfi, G.A. (1989) A combination of hypothalamic phospholipid liposomes with trazadine for treatment of depression. Acta Psychiatr. Scand. 79, 52–58.

Gibson, R.A. and Makrides, M. (2000) n-3 polyunsaturated fatty acid requirements of term infants. Am. J. Clin. Nutr. 71, 251S–255S.

Giedd, J.N. (2000) Bipolar disorder and attention-deficit/hyperactivity disorder in children and adolescents. J. Clin. Psychiatry 61, 31–34.

GISSI (1999) Dietary supplementation with n-3 polyunsaturated fatty acids and vitamin E after myocardial infarction: results of the GISSI-Prevenzione trial. Lancet 354, 447–455.

Glassman, A.H. and Shapiro, P.A. (1998) Depression and the course of coronary artery disease. Am. J. Psychiatry 155, 4–11.

Glueck, C.J., Tieger, M., Kunkel, R., Tracy, T., Speirs, J., Streicher, P. and Illig, E. (1993) Improvement in symptoms of depression and in an index of life stressors accompany treatment of severe hypertriglyceridemia. Biol. Psychiatry 34, 240–252.

Goldman-Rakic, P.S., Lidow, M.S. and Gallager, D.W. (1990) Overlap of dopaminergic, adrenergic, and serotoninergic receptors and complementarity of their subtypes in primate prefrontal cortex. J. Neurosci. 10, 2125–2138.

Goldstein, M.G. and Niaura, R. (1992) Psychological factors affecting physical condition. Cardiovascular disease literature review. Part I: Coronary artery disease and sudden death. Psychosomatics 33, 134–145.

Greene, W.A., Goldstein, S. and Moss, A.J. (1972) Psychosocial aspects of sudden death. A preliminary report. Arch. Intern. Med. 129, 725–731.

Grimsgaard, S., Bonaa, K.H., Hansen, J.B. and Myhre, E.S. (1998) Effects of highly purified eicosapentaenoic acid and docosahexaenoic acid on hemodynamics in humans. Am. J. Clin. Nutr. 68, 52–59.

Gullner, H.G. (1983) The interactions of prostaglandins with the sympathetic nervous system – a review. J. Auton. Nerv. Syst. 8, 1–12.

Hallaq, H., Smith, T.W. and Leaf, A. (1992) Modulation of dihydropyridine-sensitive calcium channels in heart. Proc. Natl. Acad. Sci. U.S.A. 89, 1760–1764.

Hamazaki, T., Sawazaki, S. and Kobayashi, M. (1996) The effect of docosahexaenoic acid on aggression in young adults. A double-blind study. J. Clin. Invest. 97, 1129–1134.

Hamazaki, T., Sawazaki, S., Nagasawa, T., Nagao, Y., Kanagawa, Y. and Yazawa, K. (1999) Administration of docosahexaenoic acid influences behavior and plasma catecholamine levels at times of psychological stress. Lipids 34, S33–S37.

Harris, W.S. (1996) Nonpharmacologic treatment of hypertriglyceridemia: focus on fish oils. Clin. Cardiol. 22, I140–I143.

Harris, W.S. (1997) n-3 fatty acids and serum lipoproteins: human studies. Am. J. Clin. Nutr. 65, 1645S–1654S.

Harris, W.S., Rambjor, G.S., Windsor, S.L. and Diederich, D. (1997) n-3 fatty acids and urinary excretion of nitric oxide metabolites in humans. Am. J. Clin. Nutr. 65, 459–464.

Hasdai, D., Scheinowitz, M., Leibovitz, E., Sclarovsky, S., Eldar, M. and Barak, V. (1996) Increased serum concentrations of interleukin-1 beta in patients with coronary artery disease. Heart 76, 24–28.

Heron, D., Shinitzky, M., Hershkowitz, M. and Samuel, D. (1980) Lipid fluidity markedly modulates the binding of serotonin to mouse brain membranes. Proc. Natl. Acad. Sci. U.S.A. 77, 7463–7467.

Hibbeln, J.R. (1998) Fish consumption and major depression. Lancet 351, 1213.

Hibbeln, J.R. (1999) Membrane lipids in relation to depression. In: M. Peet, I. Glen and D.F. Horrobin (Eds.), Phospholipid Spectrum Disorder in Psychiatry, Marius Press, Carnforth, UK, pp. 195–210.

Hibbeln, J.R. (2000) Seafood consumption and homicide mortality. World Rev. Nutr. Diet. 85, 41–46.

Hibbeln, J.R. (2001) Seafood consumption, the DHA composition of mothers' milk and prevalence of postpartum depression: a cross-national analysis. J. Affect. Disorders, in press.

Hibbeln, J.R. and Salem Jr, N. (1995) Dietary polyunsaturated fatty acids and depression: when cholesterol does not satisfy. Am. J. Clin. Nutr. 62, 1–9.

Hibbeln, J.R. and Salem Jr, N. (1996) Risks of cholesterol-lowering therapies. Biol. Psychiatry 40, 686–668.

Hibbeln, J.R., Linnoila, M., Umhau, J.C., Rawlings, R., George, D.T. and Salem Jr, N. (1998a) Essential fatty acids predict metabolites of serotonin and dopamine in cerebrospinal fluid among healthy control subjects, and early- and late-onset alcoholics. Biol. Psychiatry 44, 235–242.

Hibbeln, J.R., Umhau, J.C., Linnoila, M., George, D.T., Ragan, P.W., Shoaf, S.E., Vaughan, M.R., Rawlings, R. and Salem Jr, N. (1998b) A replication study of violent and nonviolent subjects: cerebrospinal fluid metabolites of serotonin and dopamine are predicted by plasma essential fatty acids. Biol. Psychiatry 44, 243–249.

Hibbeln, J.R., Buka, S., Yolken, R., Klebanoff, M., Majchrzak, S. and Salem Jr, N. (2000a) Maternal plasma EFA composition sampled at birth and psychotic illnesses among their children. In: Brain Reuptake and Utilization of Fatty Acids, Applications to Peroxisomal Disorders, an International Workshop, Bethesda, MD, p. 11.

Hibbeln, J.R., DePetrillo, P., Higley, J.D., Schoaf, S., Lindell, S. and Salem Jr, N. (2000b) Neuropsychiatric implications of improvements in heart-rate variability among adolescents rhesus monkeys fed formula supplemented with DHA and AA as infants. In: 4th Congress of the International Society for the Study of Lipids and Fatty Acids, Tsukuba, Japan, p. 103.

Hibbeln, J.R., Enstrom, G., Majchrzak, S., Salem Jr, N. and Traskman-Benz, L. (2000c) Suicide attempters and PUFAs: lower plasma eicosapentaenoic acid alone predicts greater psychopathology (abstract). In: 4th Congress of the International Society for the Study of Fatty Acids and Lipids, Tsukuba, Japan, p. 104.

Hibbeln, J.R., Umhau, J.C., George, D.T., Shoaf, S.E., Linnoila, M. and Salem Jr, N. (2000d) Plasma total cholesterol concentrations do not predict cerebrospinal fluid neurotransmitter metabolites: implications for the biophysical role of highly unsaturated fatty acids. Am. J. Clin. Nutr. 71, 331S–338S.

Higley, J.D. and Bennett, A.J. (1999) Central nervous system serotonin and personality as variables contributing to excessive alcohol consumption in non-human primates. Alcohol Alcohol 34, 402–418.

Higley, J.D., Hasert, M.F., Suomi, S.J. and Linnoila, M. (1991) Nonhuman primate model of alcohol abuse: effects of early experience, personality, and stress on alcohol consumption. Proc. Natl. Acad. Sci. U.S.A. 88, 7261–7265.

Hirayama, T. (1990) Life style and mortality. A large scale census-based cohort study in Japan. Contrib. Epidemiol. Biostatist. 2, Karger, New York.

Hockman, C.H., Mauck Jr, H.P. and Hoff, E.C. (1966) ECG changes resulting from cerebral stimulation. II. A spectrum of ventricular arrhythmias of sympathetic origin. Am. Heart J. 71, 695–700.

Holman, R.T., Johnson, S.B. and Ogburn, P.L. (1991) Deficiency of essential fatty acids and membrane fluidity during pregnancy and lactation. Proc. Natl. Acad. Sci. U.S.A. 88, 4835–4839.

Homayoun, P., Durand, G., Pascal, G. and Bourre, J.M. (1988) Alteration in fatty acid composition of adult rat brain capillaries and choroid plexus induced by a diet deficient in n-3 fatty acids: slow recovery after substitution with a nondeficient diet. J. Neurochem. 51, 45–48.

Hsu, J.J. and Yap, A.T. (1967) Autonomic reactions in relation to psychotropic drugs. Dis. Nerv. Syst. 28, 304–310.

Hulshoff Pol, H.E., Hoek, H.W., Susser, E., Brown, A.S., Dingemans, A., Schnack, H.G., van Haren, N.E., Pereira Ramos, L.M., Gispen-de Wied, C.C. and Kahn, R.S. (2000) Prenatal exposure to famine and brain morphology in schizophrenia. Am. J. Psychiatry 157, 1170–1172.

Ikeda, M., Yoshida, S., Busto, R., Santiso, M. and Ginsberg, M.D. (1986) Polyphosphoinositides as a probable source of brain free fatty acids accumulated at the onset of ischemia. J. Neurochem. 47, 123–32.

Ikemoto, A., Kobayashi, T., Watanabe, S. and Okuyama, H. (1997) Membrane fatty acid modifications of PC12 cells by arachidonate or docosahexaenoate affect neurite outgrowth but not norepinephrine release. Neurochem. Res. 22, 671–678.

Jones, J.P., Nicholas, H.J. and Ramsay, R.B. (1975a) Biosynthesis of cholesterol by brain tissues: distribution in subcellular fractions as a function of time after injection of [2-C-14] acetate and [U-C-14] glucose into 15 day old rats. J. Neurochem. 24, 117–121.

Jones, J.P., Nicholas, H.J. and Ramsay, R.B. (1975b) Rate of sterol formation by rat brain glia and neurons in vitro and in vivo. J. Neurochem. 24, 122–126.

Jovinge, S., Hamsten, A., Tornvall, P., Proudler, A., Bavenholm, P., Ericsson, C.G., Godsland, I., de Faire, U. and Nilsson, J. (1998) Evidence for a role of tumor necrosis factor alpha in disturbances of triglyceride and glucose metabolism predisposing to coronary heart disease. Metabolism 47, 113–118.

Kang, J.X. and Leaf, A. (1996) The cardiac antiarrhythmic effects of polyunsaturated fatty acid. Lipids 31, S41–S44.

Kannan, H., Tanaka, Y., Kunitake, T., Ueta, Y., Hayashida, Y. and Yamashita, H. (1996) Activation of sympathetic outflow by recombinant human interleukin-1 beta in conscious rats. Am. J. Physiol 270, R479–R485.

Kaplan, J.R., Manuck, S.B. and Shively, C. (1991a) The effects of fat and cholesterol on social behavior in monkeys. Psychosom. Med. 53, 634–642.

Kaplan, J.R., Pettersson, K., Manuck, S.B. and Olsson, G. (1991b) Role of sympathoadrenal medullary activation in the initiation and progression of atherosclerosis. Circulation 84, VI23–VI32.

Kaplan, J.R., Shively, C.A., Fontenot, M., Morgan, T.M., Howell, S.M., Manuck, S.B., Muldoon, M.F. and Mann, J.J. (1994) Demonstration of an association among dietary cholesterol, central serotonergic activity, and social behavior in monkeys. Psychosom. Med. 56, 479–484.

Keys, A. (1997) Coronary heart disease in seven countries. 1970 [classical article]. Nutrition 13, 250–252.

Khaykin, Y., Dorian, P., Baker, B., Shapiro, C., Sandor, P., Mironov, D., Irvine, J. and Newman, D. (1998) Autonomic correlates of antidepressant treatment using heart-rate variability analysis. Can. J. Psychiatry 43, 183–186.

Kleiger, R.E., Miller, J.P., Bigger Jr, J.T. and Moss, A.J. (1987) Decreased heart rate variability and its association with increased mortality after acute myocardial infarction. Am. J. Cardiol. 59, 256–262.

Klerman, G.L. and Weissman, M.M. (1989) Increasing rates of depression. J. Am. Med. Assoc. 261, 2229–2235.

Knapp, H.R. and FitzGerald, G.A. (1989) The antihypertensive effects of fish oil. A controlled study of polyunsaturated fatty acid supplements in essential hypertension. N. Engl. J. Med. 320, 1037–1043.

Kodavanti, U.P. and Mehendale, H.M. (1990) Cationic amphiphillic drugs and phospholipid storage disorder. Pharmacol. Rev. 42, 327–354.

Koizumi, K., Terui, N. and Kollai, M. (1985) Effect of cardiac vagal and sympathetic nerve activity on heart rate in rhythmic fluctuations. J. Auton. Nerv. Syst. 12, 251–259.

Kopin, I.J., Lake, R.C. and Ziegler, M. (1978) Plasma levels of norepinephrine. Ann. Intern. Med. 88, 671–680.

Krauss, R.M., Eckel, R.H., Howard, B., Appel, L.J., Daniels, S.R., Deckelbaum, R.J., Erdman Jr, J.W., Kris-Etherton, P., Goldberg, I.J., Kotchen, T.A., Lichtenstein, A.H., Mitch, W.E., Mullis, R., Robinson, K., Wylie-Rosett, J., Jeor, S. St., Suttie, J., Tribble, D.L. and Bazzarre, and T.L. (2000) AHA dietary guidelines:

revision 2000: A statement for healthcare professionals from the nutrition committee of the American Heart Association. Circulation 102, 2284–2299.

Kremer, J.M., Lawrence, D.A., Jubitz, W., DiGiacomo, R., Rynes, R., Bartholomew, L.E. and Sherman, M. (1990) Dietary fish oil and olive oil supplementation in patients with rheumatoid arthritis. Arthritis Rheum. 33, 810–820.

Kristensen, S.D., Schmidt, E.B. and Dyerberg, J. (1989) Dietary supplementation with n-3 polyunsaturated fatty acids and human platelet function: a review with particular emphasis on implications for cardiovascular disease. J. Intern. Med. Suppl. 225, 141–150.

Krittayaphong, R., Cascio, W.E., Light, K.C., Sheffield, D., Golden, R.N., Finkel, J.B., Glekas, G., Koch, G.G. and Sheps, D.S. (1997) Heart rate variability in patients with coronary artery disease: differences in patients with higher and lower depression scores. Psychosom. Med. 59, 231–235.

Kromhout, D. (1989) N-3 fatty acids and coronary heart disease: epidemiology from Eskimos to Western populations. J. Intern. Med. Suppl. 225, 47–51.

Kuczmierczyk, A.R., Barbee, J.G., Bologna, N.A. and Townsend, M.H. (1996) Serum cholesterol levels in patients with generalized anxiety disorder (GAD) and with GAD and comorbid major depression. Can. J. Psychiatry 41, 465–468.

Ladwig, K.H., Kieser, M., Konig, J., Breithardt, G. and Borggrefe, M. (1991) Affective disorders and survival after acute myocardial infarction. Results from the post-infarction late potential study. Eur. Heart J. 12, 959–964.

Laghrissi-Thode, F., Wagner, W.R., Pollock, B.G., Johnson, P.C. and Finkel, M.S. (1997) Elevated platelet factor 4 and beta-thromboglobulin plasma levels in depressed patients with ischemic heart disease. Biol. Psychiatry 42, 290–295.

Lands, W.E.M., Libelt, B., Morris, A., Kramer, N.C., Prewitt, T.E., Bowen, P., Schmeisser, D., Davidson, M.H. and Burns, J.H. (1992) Maintenance of lower proportions of (n-6) eicosanoid precursors in phospholipids of human plasma in response to added dietary (n-3) fatty acids. Biochim. Biophys. Acta 1180, 147–162.

Leaf, A. and Kang, J.X. (1996) Prevention of cardiac sudden death by N-3 fatty acids: a review of the evidence. J. Intern. Med. 240, 5–12.

Leaf, A. and Weber, P.C. (1987) A new era for science in nutrition. Am. J. Clin. Nutr. 45, 1048–1053.

Lechin, F., van der Dijs, B., Orozco, B., Lechin, M.E., Baez, S., Lechin, A.E., Rada, I., Acosta, E., Arocha, L. and Jimenez, V. (1995) Plasma neurotransmitters, blood pressure, and heart rate during supine resting, orthostasis, and moderate exercise conditions in major depressed patients. Biol. Psychiatry 38, 66–173.

Lehnert, H., Lombardi, F., Raeder, E., Lorenzo, A.V., Verrier, R.L., Lown, B. and Wurtman, R.J. (1987) Increased release of brain serotonin reduces vulnerability to ventricular fibrillation in the cat. J. Cardiovasc. Pharm. 10, 389–397.

Liao, D., Cai, J., Rosamond, W.D., Barnes, R.W., Hutchinson, R.G., Whitsel, E.A., Rautaharju, P. and Heiss, G. (1997) Cardiac autonomic function and incident coronary heart disease: a population-based case-cohort study. The ARIC Study. Atherosclerosis Risk in Communities Study. Am. J. Epidemiol. 145, 696–706.

Light, K., Kothandapani, R. and Allen, M.T. (1998) Enhanced cardiovascular and catecholamine responses in women with depressive symptoms. Int. J. Psychophysiol. 28, 157–166.

Linnoila, M., De Jong, J. and Virkkunen, M. (1989) Family history of alcoholism in violent offenders and impulsive fire setters. Arch. Gen. Psychiatry 46, 613–666.

Linnoila, M.V., Virkkunen, M., Scheinin, M., Nuutila, A., Rimon, R. and Goodwin, F. (1983) Low cerebrospinal fluid 5-hydroxyindoleacetic acid concentration differentiates impulsive from nonimpulsive violent behavior. Life Sci. 33, 2609–2614.

Litman, B.J., Lewis, E.N. and Levin, I.W. (1991) Packing characteristics of highly unsaturated bilayer lipids: Raman spectroscopic studies of multilamellar phosphatidylcholine dispersions. Biochemistry 30, 313–319.

Locher, R., Sachinidis, A., Steiner, A., Vogt, E. and Vetter, W. (1989) Fish oil affects phosphoinositide turnover and thromboxane A metabolism in cultured vascular muscle cells. Biochim. Biophys. Acta 1012, 279–283.

Low, P.A. (1993) Autonomic nervous system function. J. Clin. Neurophysiol. 10, 14–27.

MacNeil, B.J., Jansen, A.H., Janz, L.J., Greenberg, A.H. and Nance, D.M. (1997) Peripheral endotoxin increases splenic sympathetic nerve activity via central prostaglandin synthesis. Am. J. Physiol. 273, R609–R614.

Maes, M., Bosmans, E., Suy, E., Vandervorst, C., DeJonckheere, C. and Raus, J. (1991) Depression-related disturbances in mitogen-induced lymphocyte responses and interleukin-1 beta and soluble interleukin-2 receptor production. Acta Psychiatr. Scand. 84, 379–386.

Maes, M., Bosmans, E., Meltzer, H.Y., Scharpe, S. and Suy, E. (1993a) Interleukin-1 beta: a putative mediator of HPA axis hyperactivity in major depression? Am. J. Psychiatry 150, 1189–1193.

Maes, M., Scharpe, S., Meltzer, H.Y., Bosmans, E., Suy, E., Calabrese, J. and Cosyns, P. (1993b) Relationships between interleukin-6 activity, acute phase proteins, and function of the hypothalamic–pituitary–adrenal axis in severe depression. Psychiatry Res. 49, 11–27.

Maes, M., Meltzer, H.Y., Bosmans, E., Bergmans, R., Vandoolaeghe, E., Ranjan, R. and Desnyder, R. (1995a) Increased plasma concentrations of interleukin-6, soluble interleukin-6, soluble interleukin-2 and transferrin receptor in major depression. J. Affect. Disord. 34, 301–309.

Maes, M., Vandoolaeghe, E., Ranjan, R., Bosmans, E., Bergmans, R. and Desnyder, R. (1995b) Increased serum interleukin-1-receptor-antagonist concentrations in major depression. J. Affect. Disord. 36, 29–36.

Maes, M., Smith, R., Christophe, A., Cosyns, P., Desnyder, R. and Meltzer, H. (1996) Fatty acid composition in major depression: decreased omega 3 fractions in cholesteryl esters and increased C20:4 omega 6/C20:5 omega 3 ratio in cholesteryl esters and phospholipids. J. Affect. Disord. 38, 35–46.

Maes, M., Christophe, A., Delanghe, J., Altamura, C., Neels, H. and Meltzer, H.Y. (1999) Lowered omega-3 polyunsaturated fatty acids in serum phospholipids and cholesteryl esters of depressed patients. Psychiatry Res. 85, 275–291.

Maggioni, M., Picotti, G.B., Bondiolotti, G.P., Panerai, A., Cenacchi, T., Nobile, P. and Brambilla, F. (1990) Effects of phosphatidylserine therapy in geriatric patients with depressive disorders. Acta Psychiatr. Scand. 81, 265–270.

Malyszko, J., Urano, T., Knofler, R., Ihara, H., Shimoyama, I., Uemura, K., Takada, Y. and Takada, A. (1994) Correlations between platelet aggregation, fibrinolysis, peripheral and central serotonergic measures in subhuman primates. Atherosclerosis 110, 63–68.

Mann, J.J. (1998) The neurobiology of suicide. Nature Med. 4, 25–30.

Mann, J.J., McBride, P.A., Brown, R.P., Linnoila, M., Leon, A.C., DeMeo, M., Mieczkowski, T., Myers, J.E. and Stanley, M. (1992) Relationship between central and peripheral serotonin indexes in depressed and suicidal psychiatric inpatients. Arch. Gen. Psychiatry 49, 442–446.

Mann, J.J., Oquendo, M., Underwood, M.D. and Arango, V. (1999) The neurobiology of suicide risk: a review for the clinician. J. Clin. Psychiatry 60(Suppl. 2), 7–11.

Marangell, L.B., Zpoyan, H.A., Cress, K.K., Benisek, D. and Arterburn, L. (2000) A double blind placebo controlled study of docosahexaenoic acid (DHA) in the treatment of depression. In: 4th Congress of the International Society for the Study of Lipids and Fatty Acids, Tsukuba, Japan, p. 105. Abstract.

Martin, R.E. (1998) Docosahexaenoic acid decreases phospholipase A2 activity in the neurites/nerve growth cones of PC12 cells. J. Neurosci. Res. 54, 805–813.

Medini, L., Colli, S., Mosconi, C., Tremoli, E. and Galli, C. (1990) Diets rich in n-9, n-6 and n-3 fatty acids differentially affect the generation of inositol phosphates and of thromboxane by stimulated platelets, in the rabbit. Biochem. Pharmacol. 39, 129–133.

Melamed, S., Kushnir, T., Strauss, E. and Vigiser, D. (1997) Negative association between reported life events and cardiovascular disease risk factors in employed men: the CORDIS Study. Cardiovascular Occupational Risk Factors Determination in Israel. J. Psychosom. Res. 43, 247–258.

Mendall, M.A., Patel, P., Asante, M., Ballam, L., Morris, J., Strachan, D.P., Camm, A.J. and Northfield, T.C. (1997) Relation of serum cytokine concentrations to cardiovascular risk factors and coronary heart disease. Heart 78, 273–277.

Menotti, A., Keys, A., Aravanis, C., Blackburn, H., Dontas, A., Fidanza, F., Karvonen, M.J., Kromhout, D., Nedeljkovic, S. and Nissinen, A. (1989) Seven Countries Study. First 20-year mortality data in 12 cohorts of six countries. Ann. Med. 21, 175–179.

Merikangas, A. and Gelertner, G. (1990) Comorbidity for alcoholism and depression. Psychiatr. Clin. N. Am. 13, 613–632.

Meydani, S.N., Lichtenstein, A.H., Cornwall, S., Meydani, M., Goldin, B.R., Rasmussen, H., Dinarello, C.A. and Schaefer, E.J. (1993) Immunological effects of national cholesterol panel step-2 diets with and without fish derived n-3 fatty acid enrichment. J. Clin. Invest. 92, 105–113.

Mills, D.E. and Ward, R. (1989) Effects of n-3 and n-6 fatty acid supplementation on cardiovascular and neuroendocrine responses to stress in the rat. Nutr. Res. 9, 405–414.

Mills, D.E., Prkachin, K.M., Harvey, K.A. and Ward, R.P. (1989) Dietary fatty acid supplementation alters stress reactivity and performance in man. J. Hum. Hypertens. 3, 111–116.

Mills, D.E., Huang, Y.S., Nance, M. and Poisson, J.P. (1994) Psychosocial stress, catecholamines, and essential fatty acid metabolism in rats. Soc. Exp. Biol. Med. 205, 56–61.

Mitchell, D.C. and Litman, B.J. (1998) Effect of cholesterol on molecular order and dynamics in highly polyunsaturated phospholipid bilayers. Biophys. J. 75, 896–908.

Mitchell, D.C., Kibelbek, J. and Litman, B.J. (1992a) Effect of phosphorylation on receptor conformation: the metarhodopsin I in equilibrium with metarhodopsin II equilibrium in multiply phosphorylated rhodopsin. Biochemistry 31, 8107–8111.

Mitchell, D.C., Straume, M. and Litman, B.J. (1992b) Role of sn-2-polyunsaturated phospholipids and control of membrane receptor conformational equilibrium: Effects of cholesterol and acyl chain unsaturation on the metarhodopsin I–metarhodopsin II equilibrium. Biochemistry 31, 662–670.

Mitchell, D.C., Lawrence, J.T.R. and Litman, B.J. (1996) Primary alcohols modulate the activation of the G protein-coupled receptor rhodopsin by a lipid-mediated mechanism. J. Biol. Chem. 271, 19033–19036.

Morgan, R.E., Palinkas, L.A., Barrett-Connor, E.L. and Wingard, D.L. (1993) Plasma cholesterol and depressive symptoms in older men. Lancet 341, 75–79.

Mori, T.A., Bao, D.Q., Burke, V., Puddey, I.B. and Beilin, L.J. (1999) Docosahexaenoic acid but not eicosapentaenoic acid lowers ambulatory blood pressure and heart rate in humans. Hypertension 34, 253–260.

Morrow, J.D., Frei, B., Longmire, A.W., Gaziano, J.M., Lynch, S.M., Shyr, Y., Strauss, W.E., Oates, J.A. and Roberts, L.J. (1995) Increase in circulating products of lipid peroxidation (F2-isoprostanes) in smokers. Smoking as a cause of oxidative damage. N. Engl. J. Med. 332, 1198–1203.

Muldoon, M.F., Manuck, S.B., Mendelsohn, A.B., Kaplan, J.R. and Belle, S.H. (2001) Cholesterol reduction and non-illness mortality: meta-analysis of randomised clinical trials. Br. Med. J. 322, 11–15.

Murakami, Y., Yokotani, K., Okuma, Y. and Osumi, Y. (1996) Nitric oxide mediates central activation of sympathetic outflow induced by interleukin-1 beta in rats. Eur. J. Pharmacol. 317, 61–66.

Murray, C.J.L. and Lopez, A.D. (1996) Global Burden of Disease: A comprehensive assessment of mortality and disability from diseases, injuries and risk factors in 1990 and projected to 2020. Harvard University Press, Boston, MA.

Musselman, D.L., Tomer, A., Manatunga, A.K., Knight, B.T., Porter, M.R., Kasey, S., Marzec, U., Harker, L.A. and Nemeroff, C.B. (1996) Exaggerated platelet reactivity in major depression. Am. J. Psychiatry 153, 1313–1317.

Musselman, D.L., Evans, D.L. and Nemeroff, C.B. (1998) The relationship of depression to cardiovascular disease: epidemiology, biology, and treatment. Arch. Gen. Psychiatry 55, 580–592.

Myerburg, R.J., Kessler, K.M. and Castellanos, A. (1997) Epidemiology of sudden cardiac death: emerging strategies for risk assessment and control. In: S. Dunbar, K. Ellenbogen and A. Epstein (Eds.), Sudden Cardiac Death: Past, Present, and Future. Futura Publishing Company, Armonk, NY, pp. 29–51.

Nair, S.S., Leitch, J.W., Falconer, J. and Garg, M.L. (1997) Prevention of cardiac arrhythmia by dietary (n-3) polyunsaturated fatty acids and their mechanism of action. J. Nutr. 127, 383–393.

Nanjo, A., Kanazawa, A., Sato, K., Banno, F. and Fujimoto, K. (1999) Depletion of dietary n-3 fatty acid affects the level of cyclic AMP in rat hippocampus. J. Nutr. Sci. Vitaminol. Tokyo 45, 633–641.

Naughton, J.M., O'Dea, K. and Sinclair, A.J. (1986) Animal foods in traditional Australian aboriginal diets: polyunsaturated and low in fat. Lipids 21, 684–690.

Navarra, P., Pozzoli, G., Brunetti, L., Ragazzoni, E., Besser, M. and Grossman, A. (1992) Interleukin-1 beta and interleukin-6 specifically increase the release of prostaglandin E2 from rat hypothalamic explants in vitro. Neuroendocrinology 56, 61–68.

Nehler, M.R., Taylor Jr, L.M. and Porter, J.M. (1997) Homocysteinemia as a risk factor for atherosclerosis: a review. Cardiovasc. Surg. 5, 559–567.

Nizzo, M.C., Tegos, S., Gallamini, A., Toffano, G., Polleri, A. and Massarotti, M. (1978) Brain cortex phospholipids liposomes effects on CSF HVA, 5-HIAA and on prolactin and somatotrophin secretion in man. J. Neural Transm. 43, 93–102.

Nygard, O., Nordrehaug, J.E., Refsum, H., Ueland, P.M., Farstad, M. and Vollset, S.E. (1997) Plasma homocysteine levels and mortality in patients with coronary artery disease. N. Engl. J. Med. 337, 230–236.

Odemuyiwa, O., Poloniecki, J., Malik, M., Farrell, T., Xia, R., Staunton, A., Kulakowski, P., Ward, D. and Camm, J. (1994) Temporal influences on the prediction of postinfarction mortality by heart rate variability: a comparison with the left ventricular ejection fraction. Br. Heart J. 71, 521–527.

Ohashi, K. and Saigusa, T. (1997) Sympathetic nervous responses during cytokine-induced fever in conscious rabbits. Pflugers Arch. 433, 691–698.

Olsson, N.U., Shoaf, S. and Salem Jr, N. (1998) The effect of dietary polyunsaturated fatty acids and alcohol on neurotransmitter levels in rat brain. Nutr. Neurosci. 1, 133–140.

Olszewski, A.J. and McCully, K.S. (1993) Fish oil decreases serum homocysteine in hyperlipemic men. Coron. Artery Dis. 4, 53–60.

Olusi, S.O. and Fido, A.A. (1996) Serum lipid concentrations in patients with major depressive disorder. Biol. Psychiatry 40, 1128–1131.

Otto, S.J., Houwelingen, A.C., Antal, M., Manninen, A., Godfrey, K., Lopez-Jaramillo, P. and Hornstra, G. (1997) Maternal and neonatal essential fatty acid status in phospholipids: an international comparative study. Eur. J. Clin. Nutr. 51, 232–242.

Otto, S.J., van Houwelingen, A.C., Badart-Smook, A. and Hornstra, G. (1999) The postpartum docosahexaenoic acid status of lactating and nonlactating mothers. Lipids 34, S227.

Paez, X. and Hernandez, L. (1998) Plasma serotonin monitoring by blood microdialysis coupled to high-performance liquid chromatography with electrochemical detection in humans. J. Chromatogr. B 720, 33–38.

Pakala, R., Radcliffe, J.D. and Benedict, C.R. (1999a) Serotonin-induced endothelial cell proliferation is blocked by omega-3 fatty acids. Prostaglandins Leukotrienes. Essent. Fatty Acids 60, 115–123.

Pakala, R., Sheng, W.L. and Benedict, C.R. (1999b) Eicosapentaenoic acid and docosahexaenoic acid block serotonin-induced smooth muscle cell proliferation. Arterioscler. Thromb. Vasc. Biol. 19, 2316–2322.

Pakala, R., Sheng, W.L. and Benedict, C.R. (1999c) Serotonin fails to induce proliferation of endothelial cells preloaded with eicosapentaenoic acid and docosahexaenoic acid. Atherosclerosis 145, 137–146.

Pardridge, W.M. and Mietus, L.J. (1980) Palmitate and cholesterol transport through the blood brain barrier. J. Neurochem. 34, 463–466.

Pawlosky, R. and Salem Jr, N. (1995) Prolonged ethanol exposure causes a decrease in docosahexaenoic acid and an increase in docosapentaenoic acid in the feline brain and retina. Am. J. Clin. Nutr. 61, 1284–1289.

Pawlosky, R. and Salem Jr, N. (1998) Chronic alcohol exposure in rhesus monkeys alters electroretinograms and decreases levels of neural docosahexaenoic acid. Presented at Annual Meeting, Research Society on Alcoholism, Washington DC.

Peet, M. and Horrobin, D. (2001) Dose ranging effects of ethyl ester eicosapentaenoic acid in treatment resistant depression. Arch. Gen. Psychiatry, in press.

Peet, M., Murphy, B., Shay, J. and Horrobin, D. (1998) Depletion of omega-3 fatty acid levels in red blood cell membranes of depressive patients. Biol, Psychiatry 43, 15–319.

Peet, M., Poole, J. and Laugharne, J.D.E. (1999) Breastfeeding, Neurodevelopment and Schizophrenia. In: M. Peet, I. Glen and D.F. Horrobin (Eds.), Phospholipid Spectrum Disorder in Psychiatry, Marius Press, Carnforth, UK, pp. 159–166.

Peet, M., Brind, J., Ramchand, C.N., Shah, S. and Vankar, G.K. (2001) Two double-blind placebo-controlled pilot studies of eicosapentaenoic acid in the treatment of schizophrenia. Schizophr. Res. 49, 243–251.

Pepe, S. and McLennan, P.L. (1996) Dietary fish oil confers direct antiarrhythmic properties on the myocardium of rats. J. Nutr. 126, 34–42.

Pettegrew, J.W., Nichols, J., Minshew, N.J., Rush, A.J. and Stewart, R.M. (1982) Membrane biophysical studies of lymphocytes and erythrocytes in manic-depressive illness. J. Affect. Disord. 4, 237–247.

Philbrick, D.J., Mahadevappa, V.G., Ackman, R.G. and Holub, B.J. (1987) Ingestion of fish oil or a derived n-3 fatty acid concentrate containing eicosapentaenoic acid (EPA) affects fatty acid compositions of individual phospholipids of rat brain, sciatic nerve and retina. J. Nutr. 117, 1663–1670.

Piletz, J.E., Sarasua, M., Chotani, M., Saran, A. and Halaris, A. (1991) Relationship between membrane fluidity and adrenoreceptor binding in depression. Psychiatry Res. 38, 1–12.

Prisco, D., Filippini, M., Francalanci, I., Paniccia, R., Gensini, G.F. and Serneri, G.G. (1995) Effect of n-3 fatty acid ethyl ester supplementation on fatty acid composition of the single platelet phospholipids and on platelet functions. Metabolism 44, 562–569.

Purba, J.S., Raadsheer, F.C., Hofman, M.A., Ravid, R., Polman, C.H., Kamphorst, W. and Swaab, D.F. (1995) Increased number of corticotropin-releasing hormone expressing neurons in the hypothalamic paraventricular nucleus of patients with multiple sclerosis. Neuroendocrinology 62, 62–70.

Raadsheer, F.C., van Heerikhuize, J.J., Lucassen, P.J., Hoogendijk, W.J., Tilders, F.J. and Swaab, D.F. (1995) Corticotropin-releasing hormone mRNA levels in the paraventricular nucleus of patients with Alzheimer's disease and depression. Am. J. Psychiatry 152, 1372–1376.

Rabinowitz, S.H. and Lown, B. (1978) Central neurochemical factors related to serotonin metabolism and cardiac ventricular vulnerability for repetitive electrical activity. Am. J. Cardiol. 41, 516–522.

Raine, A., Lencz, T., Bihrle, S., LaCasse, L. and Colletti, P. (2000) Reduced prefrontal gray matter volume and reduced autonomic activity in antisocial personality disorder. Arch. Gen. Psychiatry 57, 119–127.

Refsum, H., Ueland, P.M., Nygard, O. and Vollset, S.E. (1998) Homocysteine and cardiovascular disease. Annu. Rev. Med. 49, 31–62.

Reich, P. (1985) Psychological predisposition to life-threatening arrhythmias. Annu. Rev. Med. 36, 397–405.

Robins, L.N. and Regier, D.A. (1991) Psychiatric Disorders in America: The Epidemiologic Catchment Area study. The Free Press, New York.

Roccatagliata, G., Maddini, M. and Iyaldi, M. (1978) Treatment of depression with cerebral phospholipids. (Trattamento delle depressioni con fosfolipidi cerebrali). Rass. Studi. Psichiatr. 67, 921–934.

Roche, H.M. and Gibney, M.J. (2000) Effect of long-chain n-3 polyunsaturated fatty acids on fasting and postprandial triacylglycerol metabolism. Am. J. Clin. Nutr. 71, 232S–237S.

Rosenblum, E.R., Gavaaler, J.S. and VanThiel, D.H. (1989) Lipid peroxidation: a mechanism for alcohol-induced testicular injury. Free Rad. Biol. Med. 7, 569–577.

Rousseau, D., Moreau, D., Raederstorff, D., Sergiel, J.P., Rupp, H., Muggli, R. and Grynberg, A. (1998) Is a dietary n-3 fatty acid supplement able to influence the cardiac effect of the psychological stress? Mol. Cell. Biochem. 178, 353–366.

Roy, A., Virkkunen, M. and Linnoila, M. (1987) Serotonin in suicide, violence, and alcoholism. Prog. Neuro-Psychopharmacol. Biol. Psychiatry 11, 173–177.

Rudin, D.O. (1981) The psychoses and neuroses as omega-3 essential fatty acid deficiency syndrome: substrate pellagra. Biol. Psychiatry 16, 837–850.

Rudorfer, M.V., Ross, R.J., Linnoila, M., Sherer, M.A. and Potter, W.Z. (1985) Exaggerated orthostatic responsivity of plasma norepinephrine in depression. Arch. Gen. Psychiatry 42, 1186–1192.

Rush, A.J., George, M.S., Sackeim, H.A., Marangell, L.B., Husain, M.M., Giller, C., Nahas, Z., Haines, S., Simpson, R.K. and Goodman, R. (2000) Vagus nerve stimulation (VNS) for treatment-resistant depressions: a multicenter study. Biol. Psychiatry 47, 276–286.

Russo, C., Olivieri, O., Girelli, D., Azzini, M., Stanzial, A.M., Guarini, P., Friso, S., De Franceschi, L. and Corrocher, R. (1995) Omega-3 polyunsaturated fatty acid supplements and ambulatory blood pressure monitoring parameters in patients with mild essential hypertension. J. Hypertens. 13, 1823–1826.

Salem Jr, N. (1989) Omega-3 fatty acids: molecular and biochemical aspects. In: G.A. Spiller and J. Scala (Eds.), New Roles for Selective Nutrients, Alan R. Liss, New York, pp. 109–228.

Salem Jr, N. and Ward, G. (1993) The effects of ethanol on polyunsaturated fatty acid composition. In: C. Alling, I. Diamond, S.W. Leslie, G.Y. Sun and W.G. Wood (Eds.), Alcohol, Cell Membranes, and Signal Transduction in the Brain, Plenum Press, New York, pp. 33–46.

Salem Jr, N., Serpentino, P., Puskin, J.S. and Abood, L.G. (1980) Preparation and spectroscopic characterization of molecular species of brain phosphatidylserines. Chem. Phys. Lipids 27, 289.

Salem Jr, N., Kim, H.Y. and Yergey, J.A. (1986) Docosahexaenoic acid: membrane function and metabolism, In: A. Simopoulos, R.R. Kifer and R. Martin (eds.), Health Effects of Polyunsaturated Seafoods. Academic Press, New York, pp. 263–317.

Sawazaki, S., Hamazaki, T., Yazawa, K. and Kobayashi, M. (1999) The effect of docosahexaenoic acid on plasma catecholamine concentrations and glucose tolerance during long-lasting psychological stress: a double-blind placebo-controlled study. J. Nutr. Sci. Vitaminol. Tokyo 45, 655–65.

Schomig, A., Richardt, G. and Kurz, T. (1995) Sympatho-adrenergic activation of the ischemic myocardium and its arrhythmogenic impact. Herz 20, 169–186.

Schuckit, M.A. (1986) Genetic and clinical implications of alcoholism and affective disorders. Am. J. Psychiatry 143, 140–147.

Schwartz, P.J. (1990) Cardiac sympathetic innervation and the prevention of sudden death. Cardiologia 35, 51–54.

Schwartz, P.J., La Rovere, M.T. and Vanoli, E. (1992) Autonomic nervous system and sudden cardiac death. Experimental basis and clinical observations for post-myocardial infarction risk stratification. Circulation 85, I77–I91.

Sengupta, N., Datta, S.C. and Sengupta, D.P. (1981) Platelet and erythrocyte membrane lipid and phospholipid patterns in different types of mental patients. Biochem. Med. 25, 267–275.

Severus, W.E., Ahrens, B. and Stoll, A.L. (1999) Omega-3 fatty acids – the missing link? Arch. Gen. Psychiatry 56, 380–381.

Shahar, E., Boland, L.L., Folsom, A.R., Tockman, M.S., McGovern, P.G. and Eckfeldt, J.H. (1999) Docosa-hexaenoic acid and smoking-related chronic obstructive pulmonary disease. The Atherosclerosis Risk in Communities Study Investigators. Am. J. Respir. Crit. Care Med. 159, 1780–1785.

Silvers, K. and Scott, K.M. (2000) Fish consumption and self-reported physical and mental health status. In: 4th Congress of the International Society for the Study of Lipids and Fatty Acids, Tsukuba, Japan.

Simopoulos, A. (1998) Overview of evolutionary aspects of ω-3 fatty acids in the diet. In: A. Simopoulos (Ed.), The Return of ω-3 Fatty Acids into the Food Supply. I. Land-Based Aminal Food Products and Their Health Effects, Vol. 83 of World Reviews of Nutrition and Diet, Karger, Basel, pp. 1–11.

Sinclair, A.J., O'Dea, K., Dunstan, G., Ireland, P.D. and Niall, M. (1987) Effects of plasma lipids and fatty acid composition of very low fat diets enriched with fish or kangaroo meat. Lipids 22, 53–59.

Siscovick, D.S., Raghunathan, T.E., King, I. and et, al. (1995) Dietary intake and cell membrane levels of long-chain n-3 polyunsaturated fatty acids and the risk of primary cardiac arrest. J. Am. Med. Assoc. 274, 1363–1367.

Sloan, R.P., Shapiro, P.A., Bagiella, E., Myers, M.M. and Gorman, J.M. (1999) Cardiac autonomic control buffers blood pressure variability responses. Psychosom. Med. 61, 58–68.

Sluzewska, A., Rybakowski, J., Bosmans, E., Sobieska, M., Berghmans, R., Maes, M. and Wiktorowicz, K. (1996) Indicators of immune activation in major depression. Psychiatry Res. 64, 161–167.

Smith, R.S. (1991) The macrophage theory of depression. Med. Hypotheses 35, 298–306.

Sperling, R.I., Benincaso, A.I., Knoell, C.T., Larkin, J.K., Austen, K.F. and Robinson, D.R. (1993) Dietary omega-3 polyunsaturated fatty acids inhibit phosphoinositide formation and chemotaxis in neutrophils. J. Clin. Invest. 91, 651–660.

Stanley, M., Mann, J. and Durand, G. (1983) Increased serotonin-2 binding in frontal cortex of suicide victims. Lancet 1, 214–216.

Stanley, M., Traskman-Bendz, L. and Dorovini-Zis, K. (1985) Correlations between aminergic metabolites simultaneously obtained from human CSF and brain. Life Sci. 37, 1279–1286.

Stevens, L., Zentall, S., Deck, J., Abate, M., Watkins, B., Lipp, S. and Burgess, J. (1995) Essential fatty acid metabolism in boys with attention-deficit hyperactivity disorder. Am. J. Clin. Nutr. 62(4), 761–768.

Stockert, M., Buscaglia, V. and De Robertis, E. (1989) In vivo action of phosphatidylserine, amytryptyline, and stress on the binding of [3H] imipramine to membranes of the rat cerebral cortex. Eur. J. Pharmacol. 160, 6–11.

Stoll, A.L., Locke, C.A., Marangell, L.B. and Severus, W.E. (1999a) Omega-3 fatty acids and bipolar disorder: a review. Prostaglandins Leukotrienes Essent. Fatty Acids 60, 329–337.

Stoll, A.L., Severus, W.E., Freeman, M.P., Rueter, S., Zboyan, H.A., Diamond, E., Cress, K.K. and Marangell, L.B. (1999b) Omega 3 fatty acids in bipolar disorder: a preliminary double-blind, placebo controlled trial. Arch. Gen. Psychiatry 56, 407–412.

Stordy, B.J. (1995) Benefit of docosahexaenoic acid supplements to dark adaptation in dyslexics. Lancet 346, 385.

Stordy, B.J. (2000) Dark adaptation, motor skills, docosahexaenoic acid, and dyslexia. Am. J. Clin. Nutr. 71, 323S–326S.

Susser, E.S. and Lin, S.P. (1992) Schizophrenia after prenatal exposure to the Dutch Hunger Winter of 1944–1945. Arch. Gen. Psychiatry 49, 983–988.

Tanskanen, A., Vartiainen, E., Tuomilehto, J., Viinamäki, H., Lehtonen, J. and Puska, P. (2000) High serum cholesterol and risk of suicide. Am. J. Psychiatry 157, 648–650.

Tanskanen, A., Hibbeln, J.R., Hintikka, J., Haatainen, K., Honkalampi, K. and Viinamäki, H. (2001a) Fish consumption, depression, and suicidality in a general population. Arch. Gen. Psychiatry 58, 512–513.

Tanskanen, A., Hibbeln, J.R., Tuomilehto, J., Uutela, A., Haukkala, A., Viinamäki, H., Lehtonen, J. and Vartiainen, E. (2001b) Fish consumption and depressive symptoms in the population. Psychiatr. Serv. 52, 529–531.

Terao, A., Oikawa, M. and Saito, M. (1993) Cytokine-induced change in hypothalamic norepinephrine turnover: involvement of corticotropin-releasing hormone and prostaglandins. Brain Res. 622, 257–261.

Terao, T., Yoshimura, R., Ohmori, O., Takano, T., Takahashi, N., Iwata, N., Suzuki, T. and Abe, K. (1997) Effect of serum cholesterol levels on meta-chlorophenylpiperazine-evoked neuroendocrine responses in healthy subjects. Biol. Psychiatry 41, 974–978.

Terao, T., Nakamura, J., Yoshimura, R., Ohmori, O., Takahashi, N., Kojima, H., Soeda, S., Shinkai, T., Nakano, H. and Okuno, T. (2000) Relationship between serum cholesterol levels and meta-chlorophenyl-piperazine-induced cortisol responses in healthy men and women. Psychiatry Res. 96, 167–173.

Testa, M., Yeh, M., Lee, P., Fanelli, R., Loperfido, F., Berman, J.W. and LeJemtel, T.H. (1996) Circulating levels of cytokines and their endogenous modulators in patients with mild to severe congestive heart failure due to coronary artery disease or hypertension. J. Am. Coll. Cardiol. 28, 964–971.

Thienprasert, A., Hamazaki, T., Kheovichai, K., Samuhaseneetoo, S., Nagasawa, T. and Wantanabe, S. (2000) The effect of docosahexaenoic acid on aggression/hostility in elderly subjects: A placebo-controlled double blind trial. In: 4th Congress of the International Society for the Study of Lipids and Fatty Acids, Tsukuba, Japan, p. 189. Abstract.

Tulen, J.H., Bruijn, J.A., de Man, K.J., Pepplinkhuizen, L., van den Meiracker, A.H. and Man in 't Veld, A.J. (1996) Cardiovascular variability in major depressive disorder and effects of imipramine or mirtazapine. J. Clin. Psychopharmacol. 16, 135–145.

Turnbull, A.V. and Rivier, C.L. (1999) Regulation of the hypothalamic–pituitary–adrenal axis by cytokines: actions and mechanisms of action. Physiol. Rev. 79, 1–71.

van Amsterdam, J.G. and Opperhuizen, A. (1999) Nitric oxide and biopterin in depression and stress. Psychiatry Res. 85, 33–38.

Veith, R.C., Lewis, N., Linares, O.A., Barnes, R.F., Raskind, M.A., Villacres, E.C., Murburg, M.M., Ashleigh, E.A., Castillo, S. and Peskind, E.R. (1994) Sympathetic nervous system activity in major depression. Basal and desipramine-induced alterations in plasma norepinephrine kinetics. Arch. Gen. Psychiatry 51, 411–422.

Verrier, R.L. (1987) Mechanisms of behaviorally induced arrhythmias. Circulation 76, I48–I56.

Verrier, R.L. and Mittleman, M.A. (1996) Life-threatening cardiovascular consequences of anger in patients with coronary heart disease. Cardiol. Clin. 14, 289–307.

Virkkunen, M., Rawlings, R., Tokola, R., Poland, R.E., Guidotti, A., Nemeroff, C., Bissette, G., Kalogeras, K., Karonen, S.L. and Linnoila, M. (1994) CSF biochemistries, glucose metabolism, and diurnal activity rhythms in alcoholic, violent offenders, fire setters, and healthy volunteers. Arch. Gen. Psychiatry 51, 20–27.

Virkkunen, M.E., Horrobin, D.F., Jenkins, D.K. and Manku, M.S. (1987) Plasma phospholipids, essential fatty acids and prostaglandins in alcoholic, habitually violent and impulsive offenders. Biol. Psychiatry 22, 1087–1096.

Voigt, R.G., Llorente, A., Jensen, C., Berretta, M.C., Boutte, C. and Heird, W.C. (1998) Preliminary results of a placebo controlled trial in attention deficit hyperactivity disorder. In: NIH Workshop on Omega-3 Essential Fatty Acids and Psychiatric Disorders, National Institutes of Health, Bethesda, MD, p. 24. Abstract.

von Schacky, C. and Weber, P.C. (1985) Metabolism and effects on platelet function of the purified eicosa-pentaenoic and docosahexaenoic acids in humans. J. Clin. Invest. 76, 2446–2450.

Walter, R.C., Buffler, R.T., Bruggemann, J.H., Guillaume, M.M., Berhe, S.M., Negassi, B., Libsekal, Y., Cheng, H., Edwards, R.L., von Cosel, R., Neraudeau, D. and Gagnon, M. (2000) Early human occupation of the Red Sea coast of Eritrea during the last interglacial. Nature 405, 65–69.

Weidner, G., Connor, S.L., Hollis, J.F. and Connor, W.E. (1992) Improvements in hostility and depression in relation to dietary change and cholesterol lowering. Ann. Intern. Med. 117, 820–823.

Weissman, M.M., Bland, R.C., Canino, G.J., Faravelli, C., Greenwald, S., Hwu, H.G., Joyce, P.R., Karam, E.G., Lee, C.K., Lellouch, J., Lepine, J.P., Newman, S.C., Rubio-Stipec, M., Wells, J.E., Wickramaratne, P.J., Wittchen, H. and Yeh, E.K. (1996) Cross-national epidemiology of major depression and bipolar disorder. J. Am. Med. Assoc. 276, 293–299.

Weissman, M.W. (1983) The treatment of depressive symptoms secondary to alcoholism, opiate addiction and schizophrenia: Evidence for the efficacy of tricyclics. In: P.J. Clayton and J.E. Barret (Eds.), Treatment of Depression: Old Controversies and New Approaches, Raven Press, New York, pp. 207–216.

Wickramaratne, P.J., Weissman, M.M., Leaf, P.J. and Holford, T.R. (1989) Age, period and cohort effects on the risk of major depression: results from five United States communities. J. Clin. Epidemiol. 42, 333–343.

Willatts, P., Forsyth, J.S., DiModugno, M.K., Varma, S. and Colvin, M. (1998) Effect of long-chain polyunsaturated fatty acids in infant formula on problem solving at 10 months of age. Lancet 352, 688–691.

Williams, R.B. and Littman, A.B. (1996) Psychosocial factors: role in cardiac risk and treatment strategies. Cardiol. Clin. 14, 97–104.

Willich, S.N., Maclure, M., Mittleman, M., Arntz, H. and Muller, J. (1993) Sudden cardiac death: support for a role of triggering in causation. Circulation 87, 1442–1450.

Winokur, G. (1990) The concept of secondary depression and its relationship to comorbidity. Psychiatr Clin. N. Am. 1, 567–583.

World Health Organization (1996) Fish and Fishery Products. World Apparent Consumption based on Food Balance Sheets (1961–1993), Vol. 821, Number 3. Food and Agriculture Oganization, WHO, Rome.

Yeragani, V.K., Pohl, R., Balon, R., Ramesh, C., Glitz, D., Jung, I. and Sherwood, P. (1991) Heart rate variability in patients with major depression. Psychiatry Res. 37, 35–46.

Yeragani, V.K., Pohl, R., Balon, R., Ramesh, C., Glitz, D., Weinberg, P. and Merlos, B. (1992) Effect of imipramine treatment on heart rate variability measures. Neuropsychobiology 26, 27–32.

Yeragani, V.K., Balon, R., Pohl, R. and Ramesh, C. (1995) Depression and heart rate variability. Biol. Psychiatry 38, 768–770.

Yokotani, K., Nishihara, M., Murakami, Y., Hasegawa, T., Okuma, Y. and Osumi, Y. (1995) Elevation of plasma noradrenaline levels in urethane-anaesthetized rats by activation of central prostanoid EP3 receptors. Br. J. Pharmacol. 115, 672–676.

Zimmer, L., Delion-Vancassel, S., Durand, G., Guilloteau, D., Bodard, S., Besnard, J.C. and Chalon, S. (2000a) Modification of dopamine neurotransmission in the nucleus accumbens of rats deficient in n-3 polyunsaturated fatty acids. J. Lipid Res. 41, 32–40.

Zimmer, L., Delpal, S., Guilloteau, D., Aioun, J., Durand, G. and Chalon, S. (2000b) Chronic n-3 polyunsaturated fatty acid deficiency alters dopamine vesicle density in the rat frontal cortex. Neurosci. Lett. 284, 25–28.

E.R. Skinner (Ed.), *Brain Lipids and Disorders in Biological Psychiatry*

CHAPTER 6

Plasma lipids and lipoproteins in personality disorder

E. Roy Skinner[1] and Frank M. Corrigan[2]

[1] *Department of Molecular and Cell Biology, University of Aberdeen,*
MacRobert Building, Room 809, Regent Walk, Aberdeen AB24 3FX, Scotland, UK;
E-mail: e.r.skinner@aberdeen.ac.uk, Telephone: 01224 2743194, Fax: 01224 274178,
[2] *Argyll and Bute Hospital, Lochgilphead, Argyll PA31 8LD, Scotland, UK;*
E-mail: frank.corrigan@aandb.scot, Telephone: 01546 602323, Fax: 01546 604914

1. Introduction

Brutal attacks of violence are all too frequently reported by the media throughout the world. Multiple killings of children in their schools and of adults in the market place, domestic violence, and aggression in the workplace have become sad features of modern society. In addition to the horrendous psychological effects on the survivors themselves, the emotional shock and trauma experienced by families and associates of the victims are indescribable. On the cold practical side, the economic cost of such acts, including the identification, detention and treatment, if appropriate, of the attacker, the counselling of victims and of those traumatised by the assault, together with the introduction of preventive measures to avoid further acts of violence, are enormous.

The DSMIV criteria for antisocial personality disorder stress the tendency to disregard and violate the rights of others through deceitfulness, impulsivity, aggressiveness, irresponsibility, disregard for safety and lack of remorse. These would suggest that neurobiological approaches to violence in antisocial personality disorder should focus on theory of mind abilities, including the capacity to feel remorse for suffering inflicted on others, impulsivity and aggression. There is no doubt that social factors will be important as are adverse early-life experiences (Raine et al. 1994) interacting with genetic and emotional response tendencies.

In the neurobiological approach, high-resolution brain scanning techniques such as functional magnetic resonance imaging are mapping with greater precision the areas of the brain which are associated with specific emotional responses. Such studies are assisting in identifying the brain regions where the important interactions leading to extreme behavioral responses to environmental stimuli occur.

Borderline personality disorder is a condition which is quite different clinically from anti-social personality disorder but one in which there are also problems with impulsiveness and aggression. In an attempt to enhance the understanding of this condition, Corrigan et al. (2000) have hypothesised that the tracts between the basal nuclei of the amygdala and the orbital prefrontal cortex will be particularly important in impulsiveness, while efferents from the central nucleus of the amygdala to the hypothalamus could be important in defence responses and threatening, aggressive behavior. If the affective instability is the result of conditioned or unconditional amygdalic discharges, emotionally weighted inputs to the amygdala via the lateral nucleus could be dampened by opiates while outputs via the central nucleus could be made less disturbing by benzodiazepines and ethanol; this might perhaps explain the comorbid substance abuse which is frequently found. Tracts from the basal lateral nucleus to the nucleus acumbens could be important for the sense of reward and reinforcement that gives meaning and joy to life but the positive or negative direction of the affect might depend on hemispheric lateralisation. The connections of the basal and accessory basal nuclei with the anterior cingulate cortex may be relevant not only to the experience of, but also to the verbal expressions of, distress. Thus the amygdaloid nuclei, with their projections to prefrontal, insular and cingulate cortices, as well as to hypothalamus and striatum, and their links with the thalamus for emotional processing of incoming sensory data, are crucially positioned for a role in emotional reactions and the associated behavioral responses. When these responses are rapid, and not adequately controlled prefrontally, trait impulsiveness will be high. Where anger and aggression are prominent the impulsive responses may be assaultive, physically or verbally. Thus we would expect a difference between impulsive violence and non-impulsive premeditated violence as indeed appears to be the case from genetic studies (Bergeman and Seroczynski 1998) and from neuroradiology. Raine et al. (1998) have used positron emission tomography in a study of 15 predatory murderers and 9 affective murderers. The affective murderers, who have a relatively uncontrolled and emotionally charged violence in response to aggression from others, had lower pre-frontal functioning and higher right-hemisphere sub-cortical functioning. Predatory murderers, those whose purposeful aggression is used in a cold-blooded way to achieve a goal, had pre-frontal functioning similar to that of the controls but like the affective murders had excessively high right sub-cortical activity. This exciting study suggests that the affective murderers have deficient pre-frontal regulation of excessive sub-cortical aggressive impulses while the predatory murderers have sufficient pre-frontal control to direct the sub-cortical drives to aggression. Sub-cortical areas assessed were the medial temporal lobe including the hippocampus, amygdala, thalamus and mid-brain.

There is considerable overlap between the brain regions associated with affiliation and those associated with violence. Kling and Brothers (1992) described the orbital frontal cortex, amygdala and temporal pole as being particularly important in social cognition including the perception of feelings of others, while Weiger and Bear (1988) focused on orbital frontal cortex, amygdala and temporal lobe in relation to aggression. Could violence be an extreme expression of non-affiliation, a manifestation of hate, neuroanatomically not too distinct from love? Does cruelty which involves the enjoyment of, or lack of concern for, the suffering of others require theory of mind capabilities not involved in simple predatory violence? Such questions have a degree of complexity which reduces the likelihood of answers at a simple neurochemical level. However, some enzymes exist as different isoforms in different brain regions, and their activity could be crucially altered by changes in membrane structure or by differences in concentrations of circulating factors; it would therefore seem to be important to build up knowledge of the chemical factors which can alter behavior. New et al. (1998) have focused on serotonergic inputs to the prefrontal orbital cortex and its role in the control and inhibition of aggression which may arise in subcortical or limbic nuclei such as the amygdala and the hypothalamus. The anterior cingulate cortex, which has connections with the amygdala and periaqueductal gray, is involved in the emotional weighting of internal and external stimuli to which it has access via the thalamus. Between seizures, people with epilepsy of the cingulate cortex may display sociopathic behaviors (Devinsky et al. 1995). The relative contribution to violence propensity of orbital frontal cortex, amygdala and anterior cingulate cortex has been the subject of a recent review by Davidson et al. (2000).

It is crucial to study the traits of anger, anxiety, impulsiveness and aggression in normal subjects as well as in individuals with anti-social personality disorder, a condition in which it is likely that violence is the end point of a continuum or of a number of continua involving social, environmental, nutritional and neurochemical variables. In our study of violent offenders we sought not only those who engaged in controlled premeditated violence but also those who had a low threshold for violence when angry and disinhibited. We have now gone on to look at the lipid variables associated with these traits in a normal, non-offending population as a directly opposite but, we hope, complementary approach to our earlier studies of violent offenders. Preliminary results are referred to in section 8.2 of this chapter. One of the main questions which needs to be addressed in any study of lipid variables in relation to traits predisposing to violence is why there should be greater effects in one brain region than in another. As will be seen below, apolipoproteins can have specific effects in relationship to fatty acids and neuroactive steroids, and alteration in their levels could produce subtle differences. The studies that have revealed that changes in the lipid composition of the body, especially of cholesterol and essential polyunsaturated fatty acids (EFAs), are relevant to brain function, have demonstrated not only that these are relevant in individuals with personality disorders but also in the general population.

Lipids, which comprise some 60% of the dry matter of the brain and provide the supporting membrane milieu in which the interacting functional protein components reside, may thus be susceptible to influencing the activity of regions of the brain concerned with impulsivity, aggression and other emotions. Very little information is available on the relationship between alteration in brain-lipid composition and behavioral response: further studies are urgently required on this and on the mechanisms by which specific brain lipids are transported to and taken up by the brain, which itself does not synthesise many of these components. Such studies may lead to means of rectifying defects in the composition and transport of lipids; they may thereby aid in the amelioration or prevention of the development of personality disorders and assist in the prevention of sporadic outbursts of devastating violence that can occur when precipitating factors accumulate to reach a critical level.

It is the purpose of the present chapter to review the evidence for such a hypothesis and to present the results of some recent studies carried out by the authors.

2. The composition and structure of the brain

Most of the lipid of the brain is phospholipid which forms the membrane bilayer surrounding the nerve cells and separating their contents from the external medium. Inserted in the membrane are numerous proteins, including enzymes which catalyse lipid interconversions and receptors which mediate the uptake of lipoproteins and many types of molecules. Also present are the protein systems which maintain electrochemical gradients across the membrane.

The more common type of phospholipid contains choline, ethanolamine, serine or inositol linked to the 3 position of glycerol through a phosphodiester bond, with an essential fatty acid (EFA) at the 2 position and either a saturated or unsaturated acid at the 1 position. A unique feature of the brain, along with the retina and sperm, is its amazingly high content of docosahexaenoic acid (DHA) which vastly exceeds the levels present in other body tissues. DHA and other EFAs cannot be synthesised in the human body but are formed from linoleic and linolenic acids from the diet by a series of desaturation and elongation reactions which occur mostly in the liver.

EFAs serve several very important functions in the body. First, the introduction of a double bond into the highly flexible saturated hydrocarbon chain of a fatty acid produces a degree of rigidity with restricted conformation in the chain. This, along with the presence of cholesterol, disturbs the close packing of the phospholipid molecules in the membrane-lipid bilayer and thereby causes an increase in membrane fluidity. This effect is particularly pronounced in the case of DHA where the 6 double bonds produce a highly rigid conformation (Applegate and Glomset 1991). The maintenance of the correct fatty-acid composition in the phospholipid bilayer is essential for optimal activity of membrane-bound receptors, since the latter is influenced by the fluidity of the surrounding medium. This is particularly true for the ends of the neurites, where the myriads of microconnections linking the nerve cells are responsible for the highly complex, integrated functions of the brain.

Second, several EFAs (arachidonic acid, AA; eicosapentaenoic acid, EPA; and DHA) are precursors of the eicosanoids (prostaglandins, thromboxanes and leukotrienes) which have important local physiological effects. In addition, fatty acids have been shown to influence gene expression, both directly and indirectly. Yet a further role exists for AA (and possibly other acids) in that its ethanolamide derivative (anandamide) forms an endogenous agonist of the cannabinoid recepter. This function may have considerable implications in the control of emotions and activities associated with this receptor. It is therefore apparent that the availability of the essential fatty acids and a means of their effective transport to the brain and uptake by the neurons are of paramount importance in the maintenance of normal brain function. The brain, unlike other tissues, does not use fatty acids as a source of energy to any significant extent.

3. The role of lipoproteins in lipid transport

3.1. Lipoprotein structure and metabolism

Almost all of the lipids in the blood are carried in association with the plasma lipoproteins. The chylomicrons and very-low-density lipoproteins (VLDL) are concerned largely with the transport of triglyceride, while the movement of cholesterol between body compartments is mostly carried out by the low-density lipoprotein (LDL) and high-density lipoprotein (HDL) classes. All of the lipoprotein classes consist of particles comprising a core of neutral lipid (triglyceride and cholesteryl ester) surrounded by a layer of phospholipid, unesterified cholesterol and specific apolipoproteins. The latter serve a multitude of functions, including their action as co-factors and modulators of enzymes involved in lipid metabolism and in forming ligands for membrane-bound lipoprotein receptors.

The chylomicrons, containing lipid of exogenous origin, are formed in the intestine. After removal of a large part of their triglyceride content by the action of lipoprotein lipase on the capillary walls of the extra-hepatic tissues, their remnant particles are taken up by receptor-mediated endocytosis into the liver. This is thought to occur through the LDL-related protein (LRP) receptor which recognises apolipoprotein E. The liver secretes VLDL which contains endogenous lipid, including EFAs which have been elaborated in the liver from dietary fatty-acid precursors such as linoleic and linolenic acids. The removal of triglyceride from VLDL by the action of lipoprotein lipase and hepatic lipase results in the formation of LDL, now enriched in cholesterol and containing apoB. A large proportion of this is normally taken up by the liver using the LDL receptor (specific for lipoproteins containing apoB or apoE). Modification of the LDL particle produced as a result of oxidation of the constituent polyunsaturated fatty acids leads to a loss of binding of this lipoprotein to the hepatic LDL receptors but creates a strong ligand for the binding to scavenger receptors ("oxidised LDL receptors") present in many tissues. This produces an elevation in the level of plasma cholesterol and an increased rate of uptake by peripheral tissues including artery walls, thereby promoting atherosclerosis (Parthasarathy and Rankin 1992). This process is opposed by the action of HDL (Skinner 1994).

HDL serves an important function, *inter alia,* in reverse cholesterol transport whereby, in association with lecithin:cholesterol acyltransferase, it takes up cholesterol effluxed from peripheral cells, including those of the artery wall, and transports it to the liver where it is thought to be taken up by the scavenger receptor, SR-BI. It is then excreted from the body in the form of bile salts. HDL therefore provides an anti-atherogenic effect.

The mechanism by which circulating fatty acids and cholesterol are transported to and taken up by the brain is not clearly understood at the present time. Before considering current ideas on this topic, we will examine the lipoprotein apparatus that is known to be present in the brain.

ApoE, ApoA-IV, apoJ and apoD have been shown to be present in the brain. While low concentrations of several other apolipoproteins with the exception of apoB have been found in the brain, these may arise through filtration from the plasma; messenger RNAs coding for them could not be detected (Beffert et al. 1998).

3.2. Apolipoprotein E

ApoE is synthesised by astrocytes and microglia but it remains uncertain whether its presence in neurons is due to synthesis by these cells or is a result of the uptake of apoE synthesised in the astrocytes. The rate of synthesis of apoE is increased following nerve damage (Ignatius et al. 1986). Strong evidence has been provided for a role of apoE in the transport of cholesterol between degrading nerve cells and regenerating neurons. Cholesterol liberated from the degrading cells may combine with newly synthesed apoE to form apoE–cholesterol–lipoprotein complexes. These are released into the circulation and taken up by local neurons undergoing axonal growth or synaptogenesis via apoE receptors (see below). After internalisation, the cholesterol is released and utilised in the formation of new membranes, while the synthesis of cholesterol is repressed through the downregulation of HMG-CoA reductase.

ApoE occurs as 3 isoforms of the protein due to the mutation of cysteine to arginine residues at positions 112 and 158; they are designated apoE2, apoE3 and apoE4, each encoded by a distinct gene. A high frequency of the apoE4 allele is present in subjects with late-onset Alzheimer's disease as discussed elsewhere in this volume.

It has been demonstrated that the addition of apoE3 to rabbit dorsal root ganglion neurons in culture produces a very significant increase in neurite extension, whereas this was markedly decreased when apoE4 was added. This effect was observed only when apoE was presented in the form of cholesterol-rich lipoprotein (e.g. β-VLDL) and was demonstrated to be taken up into the neurons by a lipoprotein receptor-mediated process. It is proposed that stimulation of neurite extension by apoE in vivo might underly the process of nerve regeneration or the formation of synaptic connections during neural reassembly.

Another function of apoE lies in its association with tau protein. The differential binding of the different isoforms of apoE to tau protein modulates the phosphorylation of the latter and thereby influences the association of these units in the formation of microtubules and thereby the stabilisation of the neuronal cytoskeleton. Furthermore, the increased deposition of β-amyloid in the cortex of patients with Alzheimer's disease who carry the apoE4 allele may be a reflection of the more rapid binding of apoE4 to β-amyloid protein than apoE3. These findings are of obvious relevance to the formation of the neurofibrillary tangles and amyloid plaques associated with this disease.

3.3. Apolipoprotein D

This is a member of the lipocalin family of proteins and was originally identified as a component of HDL. It has a low affinity for cholesterol, but binds arachidonic acid, progesterone and pregnenolone with high affinity. It is reported to be synthesised by astrocytes and oligodendrocytes (and possibly neurons) in the hippocampus, its rate of synthesis being significantly increased following neural damage. ApoD therefore appears to have a specific role in nerve regeneration with possible implications with the binding and transport of arachidonic acid and some sterol hormones (Terrisse et al. 1998). The binding of apoD to neurosteroids is particularly interesting in this context because of behavioral changes which can occur in women during the menstrual cycle and because of the vast over-representation of men as the perpetrators of violent acts (The Demonic Male).

3.4. Apolipoprotein A-IV

This apolipoprotein is a major component of newly secreted chylomicrons and of some subspecies of HDL. Strong evidence has been supplied for a role in cholesterol transport, since its addition to human skin fibroblasts in culture significantly increases the rate of cholesterol efflux from these cells (Stein et al. 1986). ApoA-IV is present, albeit in low concentrations relative to plasma, in the brain and CSF. Its synthesis in the CNS could not be demonstrated by several groups of investigators; its presence in the brain is assumed to be due to diffusion or transport from the plasma.

3.5. Apolipoprotein J

ApoJ, alternatively called clusterin, also serves a role in lipid transport but in addition is involved in other functions including cell adhesion and complement inhibition. The distribution of messenger RNA for apoJ demonstrates that the synthesis of this apolipoprotein occurs in astrocytes, but not in microglial nor endothelial cells. The observation that relatively high concentrations of apoJ messenger RNA are present near to the vasculature may suggest that apoJ is involved in lipid transport between the circulation and the CNS, analogous to its proposed role with respect to other tissues. As with apoD and apoE, the synthesis of apoJ is increased with brain injury, though it is also increased in patients with conditions associated with lipid disorders such as retinitis pigmentosa and Pick's disease.

4. The nature of lipoprotein particles in the CNS and their uptake by neuronal cells

Practical difficulties of considerable magnitude exist in the isolation and characterization of lipoprotein particles in their native form from brain tissues. The disruption of brain cells for the release of these particles is likely to produce changes in the particles themselves, while ultracentrifugation and possibly other methods required for separation of the particles according to density and size produce significant dissociation of apolipoproteins from the lipoprotein particles. Nevertheless, the presence of HDL-sized particles in the brain containing apoE or in the CSF containing apoE and apoA-I (Chiba 1991) was reported. More recent studies, however, have shown that cultured astrocytes grown in plasma-free medium secrete lipoprotein particles containing little core lipid, are mainly discoidal in shape and contain apoE and apoJ, whereas CSF lipoproteins have a size, density and lipid composition resembling plasma HDL and contain apoA-I and apoA-II in addition to apoE and apoJ (LaDu et al. 1998).

Strong evidence has been provided for the existence of a lipoprotein receptor-mediated mechanism for the uptake of cholesterol by neuronal cells (Posse de Chaves et al. 1997). When the cells were incubated with pravastatin, cholesterol synthesis was inhibited and axonal growth impaired. The observation that normal axonal growth was restored on addition of cholesterol in the form of lipoprotein supports the concept that lipoproproteins can be taken up by the axons from the surrounding microenvironment to supply sufficient cholesterol for axonal growth. The effect was restored with the addition of LDL or HDL_2 (which do not contain apoE) or with HDL_3 (which does contain apoE). Whereas normal growth of pravastatin-treated cells occurred when LDL was added to the separated cell bodies, HDL_2 and HDL_3 were effective only with preparations of distal axons. It is suggested that LDL and HDL may be taken up by different mechanisms in different neuronal regions with differential expression of the LDL receptor, the LRP receptor and the scavenger receptor, SR-IB. All of these receptors have been reported to be present in neuronal cells, though their distribution in the neurons does not appear to have been investigated. Similar experiments in which the synthesis of phospholipids by neurons was inhibited failed to support the concept that phospholipids are also taken up by a lipoprotein-mediated process.

5. The cholesterol hypothesis of violence and personality disorder

In this section, we will discuss the evidence for and against the hypothesis that low or lowered levels of plasma cholesterol are associated with an increased risk of death from suicide, accident or violence, a proposition that has aroused considerable controversy.

Attention was first focussed on the possibility of such a relationship by the results of a series of randomized primary trials aimed at determining the effect of lowering cholesterol concentration on reducing the risk of coronary heart disease. While these trials established that lowering of cholesterol in large groups of middle-aged men with primary hypercholesterolaemia was associated with reduced incidence of coronary heart events, total mortality was not significantly reduced. This failure arose in part from a marked increase in the number of deaths from violence, suicide and accidents observed in each case.

A meta-analysis of six primary prevention trials of cholesterol reduction by either dietary or pharmacological means (Muldoon et al. 1990) showed that while cholesterol reduction tends to lower mortality from coronary heart disease, total mortality was not affected by treatment and there was a significant increase in deaths not related to illness (violence, suicide and accidents). This analysis showed that while neither dietary nor drug treatment influenced total mortality, cholesterol reduction by pharmacological, but not dietary treatment, modestly lowered mortality from coronary heart disease yet significantly increased deaths from violence, suicide and accidents.

A further meta-analysis of more recent randomized trials using HMG-CoA reductase inhibitors, however, found a significant reduction in both cardiovascular and total mortality, with no increase in deaths due to violent causes (Herbert et al. 1997). Patient selection may have been a factor causing this difference since subjects with a prior risk of violence such as a history of alcohol abuse or psychiatric illness were excluded from the study. Also, benefits of statins other than cholesterol reduction may be a further cause of the differences observed with nonstatin drugs. Further research is needed to understand the effect of these agents on violence, suicide and accidents. In addition, the studies used in these analyses have been aimed at high coronary-risk subjects with primary hypercholesterolaemia, and the results obtained may not necessarily be applicable to alterations in cholesterol levels in normocholesterolaemic individuals since these groups of individuals may handle cholesterol differently.

The investigation of a direct relationship between plasma cholesterol concentration and impulsive behavior has been the subject of a number of reports. These have demonstrated an association between plasma cholesterol concentration and antisocial personality disorder (Virkkunen 1979, Freedman et al. 1987), habitual violence in homicidal offenders (Virkkunen 1983), and aggressive conduct disorders in children and adolescents with attention-deficit disorder (Virkkunen and Penttinen 1984). Recently, patients with borderline-personality disorder were found to have significantly lower serum cholesterol levels than non-borderline personality disorders, with male patients having lower cholesterol levels than female patients (New et al. 1999). However, other studies have failed to find a relationship between cholesterol concentration and aggressive behavior in male patients with antisocial personality disorder (Stewart and Stewart 1981) or between the concentration of cholesterol or any of its component lipoprotein fractions in individuals convicted of violent crimes and non-violent control subjects (Gray et al. 1993).

Evidence has also been provided which suggests that low baseline levels of plasma cholesterol *per se,* as opposed to the lowering of cholesterol levels, are associated with death from trauma or suicide (Lindberg et al. 1992, Muldoon et al. 1993, Sullivan et al. 1994, Kunugi et al. 1997, Maes et al. 1997, Alvarez et al. 1999), though this relationship has not been replicated in some other studies (Pekkanen et al. 1989, Markovitz et al. 1997). Finally, animal studies have demonstrated that male monkeys assigned to a low fat/low cholesterol diet showed more aggressive behavior than monkeys on a high fat/high cholesterol diet (Kaplan et al. 1991, 1994).

A recent paper by Golomb (1998) provides a systematic review of the literature on low or lowered cholesterol concentrations and violent death which meets a set of inclusion criteria that permit a causal connection to be evaluated. This study concluded that a significant association between low or lowered cholesterol levels and violence occurred across many types of studies with the data on this association conforming to the criteria for a causal association, thus raising concerns in risk–benefit analyses for cholesterol screening and treatment. These conclusions have been criticised on the basis that the analysis was biased against trials which used statins (see above) and that increased aggression may arise through hunger for which low cholesterol can be a marker; in many trials, however, the treatment and placebo groups received identical diets, while primates fed cholesterol-lowering diets show increased violence in spite of *ad libitum* feeding and with no weight loss.

A more recent analysis has shown that a modest increase in non-illness mortality may occur with dietary interventions and non-statin drugs (Muldoon et al. 2001).

6. Lipid alterations in individuals with personality disorder

Since the brain contains relatively large amounts of EFAs which are vital components of the diet or are produced from the latter by metabolic modification, it is perhaps not surprising that alteration in their concentrations is associated with several psychiatric disorders. Accordingly, reduced levels of n-6 fatty acids in the plasma phospholipids and in the frontal cortex of the brain have been observed in schizophrenic patients (Horrobin et al. 1989, 1991). In alcoholics, cognitive performance which has been impaired by the effect of alcohol is improved by dietary supplementation with n-6 fatty acids (Skinner et al. 1989), while in Alzheimer's disease, differences in the fatty-acid composition of different regions of the brain (Tilvis et al. 1987, Skinner et al. 1993) and of the plasma phospholipids (Corrigan et al. 1991) have been observed. Changes in EFAs and cholesterol in depression are discussed in chapter 5 in this volume.

Considerable interest is now emerging on the possible association between fatty-acid levels and aggressive behavior. Virkkunen et al. (1987) reported abnormalities of plasma phospholipid fatty acids in violent offenders, which included an increased concentration of dihomogammalinolenic acid with a lowering of arachidonic acid and reduced levels of docosahexaenoic acid.

We have studied the fatty acids of red-cell phospholipids, plasma phospholipids, cholesteryl esters and high-density lipoproteins of 19 violent offenders and 25 age-matched control subjects (Corrigan et al. 1994). There were significantly higher levels of oleic acid (18:1 n-9) in plasma phospholipids, cholesteryl esters and high-density lipoproteins of the offenders than of the control subjects, and lower concentrations of arachidonic acid (20:4 n-6) in the same lipids in the offenders. There were also decreased concentrations of other n-6 and n-3 long-chain polyunsaturated fatty acids in the offenders. The reductions in n-6 fatty acids were evident without an increase in linoleic acid (18:2 n-6), indicating that it is unlikely that the findings can be explained by altered activity of the delta-6 desaturase enzyme.

In a recent well-controlled study on a group of subjects selected for their history of violent, impulsive behaviors and nonviolent control subjects, Hibbeln et al. (1998) showed that the violent group had significantly lower concentrations of cerebrospinal fluid 5-hydroxyindoleacetic acid (CSF 5-HIAA), a metabolite that reflects serotonin turnover predominantly in the frontal cortex, than the nonviolent subjects. The observation that plasma docosahexaenoic acid was negatively correlated with CSF 5-HIAA only among violent subjects suggests that dietary essential fatty acids may change neurotransmitter concentrations and could alter impulsive and violent behavior. Another exciting possibility is that there is a transport or metabolic defect in the violent offender which alters the relationship between essential-fatty-acid intake in the diet and brain concentrations of the fatty acids with subsequent effects on neurotransmitter function. Identification of an intrinsic, presumably genetically determined, difference in lipid metabolism which predisposes to involvement in violent crime would be a major advance. The authors suggest that these findings, which replicate a previous study, are consistent with the hypothesis that low plasma DHA concentration, rather than low plasma cholesterol concentration, may increase predisposition to hostility and depression. Well-controlled clinical trials are clearly needed.

7. Biological mechanisms relating alterations in lipid composition with aggression and violence

The above discussion has provided some tantalizing evidence for the hypothesis that low or lowered cholesterol and alteration in fatty-acid composition are associated with violence, though further well-controlled studies are needed to fully test the validity of the relationship. Several mechanisms have been proposed which may help to explain the link between lipids and psychiatric disorders.

The activity of brain cell-surface receptors and other proteins, including enzymes, and ion channels is affected by the fluidity of the regions of the membrane in which the receptor is embedded, and this in turn is influenced by the lipid composition. The insertion of cholesterol into the phospholipid bilayer leads to a decrease in fluidity, while the presence of long-chain polyunsaturated fatty acids increases the fluidity due to the rigidity and conformational change introduced into these molecules by the introduction of double bonds (see section 2).

As early as 1980, Heron et al. demonstrated that when the cholesterol content of mouse synaptic brain membranes was increased, thereby increasing the microviscosity, there was a five-fold increase in the specific binding of serotonin. Since membrane-associated cholesterol freely exchanges with serum cholesterol, it was hypothesised that a reduction in plasma cholesterol may decrease the brain-cell membrane cholesterol, lower the microviscosity, and thereby decrease the uptake of serotonin into the brain cells. In view of the relationship between low levels of serotonin and violence and the role of serotonin in impulse control, this hypothesis would explain the relationship between low plasma cholesterol and violence, suicide and accidents (see Engelberg 1992).

More recently, attention has focused on the role of essential polyunsaturated fatty acids in influencing the activity of brain receptors. It has been suggested that dietary modification aimed at cholesterol reduction may have the effect of reducing the level of n-3 fatty acids in the tissues (Sinclair et al. 1987). Low cholesterol could therefore possibly represent a marker for n-3 fatty-acid deficiency.

Polyunsaturated fatty acids, rather than cholesterol, have been shown to have an effect on cell signalling cascades, including those systems mediated by G-protein transduction. For example, in rats fed a diet enriched with n-3 fatty acids, there was increased coupling of G_s to adenyl cylase stimulated by glucagon as compared with rats on a diet enriched with n-6 fatty acids (Ahmad et al. 1989). It has been demonstrated that, at least for some systems, the essential-fatty-acid composition has an important influence on G-protein-mediated signal transduction, whereas cholesterol plays only a minor role. An outcome of many of these signalling events is the release of fatty acids from membrane phospholipids, and these may act as signalling molecules themselves or may be converted by the action of appropriate enzymes to prostaglandins, thromboxanes and leukotrienes (eicosanoids) which influence a host of physiological processes. Thus small alterations in the fatty-acid composition of brain-membrane phospholipids may affect signal transduction and lead to changes in brain function and behavior (Hibbeln and Salem 1995).

Fatty acids also serve roles in gene expression, both directly through specific interactions with binding proteins that interact with elements in the 5'-flanking region of target genes, and indirectly through modulation of cell signalling pathways.

A further consideration is whether violent subjects may have a defect in the system by which essential polyunsaturated fatty acids are transported to and taken up by the brain. Linoleic and linolenic acids, essential acids of dietary origin, are transported by the chylomicrons largely to the liver. Although the brain contains the desaturase and elongase enzymes required for their conversion to other essential acids such as DHA, AA and EPA, these interconversions are effected mainly in the liver. The mechanisms by which these fatty acids are transported to the brain, however, appears to be poorly understood at the present time. While serum albumin forms complexes with free fatty acids and transports them to the liver, it has not been established that it provides a major route of transport of essential fatty acids to the brain. Areas of the brain contain lipoprotein lipase, though its substrate, VLDL, is too large to cross the blood–brain barrier. This is not the case with some of the smaller subfractions of HDL which is a possible candidate for the transport of essential fatty acids from the liver to the brain. Such a role for HDL is also supported by the fact that its fatty-acid content is in continual exchange with that of other lipoprotein fractions through the action of cholesterol ester transfer protein, and its large content of phospholipid has a high turnover rate and exchanges rapidly with membrane phospholipids. Furthermore, the brain contains a number of cell-surface receptors with high affinity for HDL apolipoproteins such as apolipoprotein E. Alterations in both the lipid and the apolipoprotein composition of HDL in violent subjects are therefore possibly worthy of investigation. The results of such a study are described below.

8. Plasma lipoproteins and apolipoproteins in violent offenders and in normal individuals

The following two studies were performed with the view to identifying the changes in plasma lipoproteins and apolipoproteins which underlie the changes in total plasma cholesterol observed in some investigations on violent subjects.

8.1. Violent offenders

Concentrations of plasma lipoproteins were measured in 15 men serving prison sentences for offences involving violence, and 25 well-matched control male subjects. In addition, the levels of HDL subfractions (see Skinner 1994) and of apolipoproteins associated with the total HDL fraction were determined as these were considered to be the components most likely responsible for the transport of lipids to the brain (Gray et al. 1993). There were no significant differences in total plasma cholesterol or in any of the major lipoprotein fractions, and the distribution of HDL subfractions (HDL_{2b}, HDL_{2a}, HDL_{3a}, HDL_{3b}, HDL_{3c}), separated according to particle size by gradient gel electrophoresis, did not differ between the offenders and the control subjects. ApoE and apoA-IV, measured as percentages of total apolipoproteins of HDL, were, however, significantly higher in the offenders than in the control group. As discussed previously in section 3, both of these apolipoproteins are implicated in the transort of cholesterol between tissues. Since these two apolipoproteins are expressed by genes on different chromosomes, it is unlikely that they are both markers of a genetic predisposition to violence, and it is suggested that the increased levels of these apolipoproteins could be part of a homeostatic mechanism for maintaining cholesterol concentrations at a level better suited to overall functioning and to overcome a tendency in these individuals to have low cholesterol levels.

8.2. Plasma lipids and personality traits in a normal population

On the assumption that the violence of offenders is at the extreme end of a spectrum of aggression and impulsiveness we have studied these traits in relation to plasma lipids in a group of non-offending volunteers who have completed the Special Hospitals Assessment of Personality Scale (SHAPS, Blackburn 1982) (Corrigan, F.M., Skinner, E.R., Lyon, C. and Stewart, E.C., unpublished data). This scale gives measures of anxiety, extraversion, hostility, introversion, depression, tension, impulsiveness, aggression, lying and psychopathy. Lipoprotein fractions were separated by ultracentrifugation at increasing solvent density, and apolipoproteins were determined by immuno-assay techniques. In men, but not in women, the relationship between the concentration of apoE associated with HDL and the SHAPS psychopathic deviate (PD) score ($n = 43$, RS $= 0.27$, $p < 0.1$) did not quite reach statistical significance, but when the subjects were divided into low- and high-apoE groups (<1.35 and >1.35 mg per 100 ml plasma), the difference in the PD score between the two groups was significant (Low ApoE, $n = 27$: PD 15.07 ± 3.40 vs. High ApoE, $n = 16$: PD 18.19 ± 3.87, $p < 0.009$). There was no association of the HDL apoE concentration with the measures of aggression or impulsiveness. The results of this study support our previous findings described above of a relationship between HDL apoE and personality disorder, the correlation between the two parameters in the male volunteers being positive. ApoE-deficient mice develop hyperlipidaemia, atherosclerosis and central nervous system neurodegeneration, and Mato et al. (1999) have demonstrated regional differences in the brain lipids of these mice. Differences in the levels of HDL apoE in personality disorders and in relation to degree of psychopathic deviate scoring in non-offending populations may therefore be a reflection of impaired transport of fatty acids by a subfraction of HDL rich in apoE rather than, or in addition to, the homeostatic mechanisms for the maintenance of cholesterol levels.

9. Summary and conclusions

This chapter has summarized epidemiologal, clinical and biochemical evidence for a relationship between the concentration of cholesterol and EFAs and violence. Many, but not all randomized trials on the effect of cholesterol lowering have shown an increase in the incidence of deaths due to violent causes, the outcome depending on whether statin or non-statin drugs were employed. Other studies have shown that violent behavior is associated with low cholesterol per se. The concentrations of specific EFAs have also been shown to be altered in some subjects with personality disorders in a number of studies. Mechanisms which have been proposed to explain these relationships include alteration in serotonin receptor activity caused by a change in membrane microviscosity, and the effect of EFAs on cell-signalling processes and on the production of eicosanoids. It has been suggested that the association between lipids and violence may lie in alterations in EFA composition rather than in cholesterol level, the latter possibly being a reflection of dietary changes in n-3 and n-6 EFAs.

Our previous observations of altered HDL apolipoproteins in violent offenders support the hypothesis that an abnormality of lipid transport to the brain may be present in those especially predisposed to violence. We have argued that there will be a continuum of susceptibility to eruption of subcortical anger through prefrontal impulse control defences, and that this balance depends on many social, environmental, nutritional and neurochemical factors. A shift along the continuum to a point at which the behavior becomes overt and dangerous can occur with comparatively small modifications in the lipid structure of the brain. The involvement of apolipoproteins in the transport of cholesterol and fatty acids to and within the brain is an area in which the magnitude of change commonly induced by cholesterol-lowering drugs is clearly sufficient, by whatever mechanisms, to induce changes in neurotransmitter function that can have a major impact on the life of a person and on those around them.

These observations provide intriguing evidence for, but do not prove, the hypothesis that cholesterol and EFA levels are associated with violent behavior. Further proof for such a relationship and the possibility of a causal relationship must await the results of further studies.

References

Ahmad, S.N., Alma, B.S. and Alam, S.Q. (1989) Dietary omega-3 fatty acids increase guaninenucleotide binding proteins and adenyl cyclase activity in rat salivary glands. FASEB J. 3, a948 (abstract 4196).

Alvarez, J.C., Cremniter, D., Lesieur, P., Gregoire, A., Gilton, A., Macquinmavier, I., Jarreau, C. and Spreuxvaroquaux, O. (1999) Low blood cholesterol and low platelet serotonin levels in violent suicide attempters. Biol. Psychiatry 45, 1066–1069.

Applegate, K.R. and Glomset, J.A. (1991) Effect of acyl chain unsaturation on the conformation of model diacylglycerols: a computer modeling study. J. Lipid Res. 32, 1635–1644.

Beffert, U., Danik, M., Krzywkowski, P., Ramassamy, C., Berrada, F. and Poirier, J. (1998) The neurobiology of apolipoproteins and their receptors in the CNS and Alzheimer's disease. Brain Res. Rev. 27, 119–142.

Bergeman, C.S. and Seroczynski, A.D. (1998) Genetic and environmental influences on aggression and impulsivity. In: M. Maes and E.F. Coccaro (Eds.), Neurobiology and Clinical Views on Aggression and Impulsivity, Vol. 5, Wiley, New York, pp. 63–80.

Blackburn, R. (1982) The special hospitals assessment of personality and socialization. Unpublished manuscript (personal communication), Park Lane Hospital, Liverpool.

Chiba, H. (1991) Apolipoproteins in rat cerebrospinal fluid: a comparison with plasma lipoprotein metabolism and effect of aging. Neurosci. Lett. 133, 207–210.

Corrigan, F.M., van Rhijn, A.G., Ijomah, G., McIntyre, G., Skinner, E.R., Horrobin, D.F. and Ward, N.I. (1991) Tin and fatty acids in dementia. Prostaglandins, Leukotrienes Essent. Fatty Acids 43, 229–238.

Corrigan, F.M., Gray, R., Strathdee, A., Skinner, R., van Rhijn, A. and Horrobin, D.F. (1994) Fatty acid analysis of blood from violent offenders. J. Forens. Psychiatry 5, 83–92.

Corrigan, F.M., Davidson, A. and Heard, H. (2000) The role of dysregulated amygdalic emotion in borderline personality disorder. Med. Hypotheses 54, 574–579.

Davidson, R.J., Putnam, K.M. and Larson, C.L. (2000) Dysfunction in the neural circuitry of emotional regulation – a possible prelude to violence. Science 289, 591–594.

Devinsky, O., Morrell, M.J. and Vogt, B.A. (1995) Contributions of anterior cingulate cortex to behaviour. Brain 118, 279–306.

Engelberg, H. (1992) Low serum cholesterol and suicide. Lancet 339, 727–729.

Freedman, D.S., Byers, T., Barrett, D.H., Stroup, N.E., Eaker, E. and Monroe-Blum, H. (1987) Plasma lipid levels and psychologic characteristics in men. Am. J. Epidemiol. 141, 507–517.

Golomb, B.A. (1998) Cholesterol and violence: is there a connection? Ann. Intern. Med. 128, 478–486.

Gray, R.F., Corrigan, F.M., Strathdee, A., Skinner, E.R., vanRhijn, A.G. and Horrobin, D.F. (1993) Cholesterol metabolism and violence: a study of individuals convicted of violent crimes. NeuroReport 4, 754–756.

Herbert, P.R., Gaziano, J.M., Chan, K.S. and Hennekens, C.H. (1997) Cholesterol lowering with statin drugs, risk of stroke, and total mortality. J. Am. Med. Assoc. 278, 313–321.

Heron, D.S., Shinitzky, M., Hershowitz, M. and Samuel, D. (1980) Lipid fluidity markedly modulates the binding of serotonin to mouse brain membranes. Proc. Natl. Acad. Sci. U.S.A. 77, 7463–7467.

Hibbeln, J.R. and Salem Jr, N. (1995) Dietary polyunsaturated fatty acids and depression: when cholesterol does not satisfy. Am. J. Clin. Nutr. 62, 1–9.

Hibbeln, J.R., Linnoila, M., Umhau, J.C., Rawlings, R., George, D.T. and Salem Jr, N. (1998) Essential fatty acids predict metabolites of serotonin and dopamine in cerebrospinal fluid among healthy control subjects and early- and late-onset alcoholics. Biol. Psychiatry 44, 235–242.

Horrobin, D.F., Manku, M.S., Morse-Fisher, N., Vaddadi, K.S., Courtney, P., Glen, A., Glen, I.M., Spellman, M. and Bates, C. (1989) Essential fatty acids in plasma phospholipids in schizophrenics. Biol. Psychiatry 25, 562–568.

Horrobin, D.F., Manku, M.S., Hillman, H. and Glen, A.I.M. (1991) Fatty acid levels in the brains of schizophrenics and normal controls. Biol. Psychiatry 30, 795–805.

Ignatius, M.J., Gebicke-Harter, P.J., Pate Skene, J.H., Schilling, J.W., Weisgraber, K.H., Mahley, R.W. and Shooter, E.M. (1986) Expression of apolipoprotein E during nerve degeneration and regeneration. Proc. Natl. Acad. Sci. U.S.A. 83, 1125–1129.

Kaplan, J.R., Manuck, S.B. and Shively, C. (1991) The effects of fat and cholesterol on social behavior in monkeys. Psychosom. Med. 53, 634–642.

Kaplan, J.R., Shively, C.A., Fontenot, M.B., Morgan, T.M., Howell, S.M., Manuck, S.B., Muldoon, M.F. and Mann, J.J. (1994) Demonstration of an association among dietary cholesterol, central serotonergic activity, and social behavior in monkeys. Psychosom. Med. 56, 479–484.

Kling, A.S. and Brothers, L.A. (1992) The amygdala and social behavior. In: J.P. Aggleton (Ed.), The Amygdala: Neurobiological Aspects of Emotion, Memory, and Mental Dysfunction, Wiley, New York, pp. 353–377.

Kunugi, H., Takei, N., Aoki, H. and Nanko, S. (1997) Low serum cholesterol in suicide attempters. Biol. Psychiatry 41, 196–200.

LaDu, M.J., Gilligan, S.M., Lukens, J.R., Cabana, V.G., Reardon, C.A., Van Eldik, L.J. and Holzman, D.M. (1998) Nascent astrocyte particles differ from lipoproteins in CSF. J. Neurochem. 70, 2070–2081.

Lindberg, G., Rastam, L., Gullberg, B. and Eklund, G.A. (1992) Low serum cholesterol concentration and short term mortality from injuries in men and women. Br. Med. J. 305, 277–279.

Maes, M., Smith, R., Christophe, A., Vandoolaegh, E., van Gastel, A., Neels, H., Demedts, P., Wauters, A. and Meltzer, H. (1997) Lower serum high density lipoprotein cholesterol (HDL-C) in major depression and in depressed men with serious suicidal attempts: relationship with immune-inflammatory markers. Acta Psychiatr. Scand. 95, 212–221.

Markovitz, J.H., Smith, D., Racynski, J.M., Oberman, A., Williams, O.D., Knox, S. and Jacobs, D.R. (1997) Lack of relations of hostility, negative affect and high risk behavior with low plasma lipid levels in the Coronary Artery Risk Development in Young Adults Study. Arch. Intern. Med. 157, 1953–1959.

Mato, M., Ookawara, S., Mashiko, T., Sakamoto, A., Mato, T.K., Maeda, N. and Kodama, T. (1999) Regional difference of lipid distribution in brain of apolipoprotein E deficient mice. Anat. Rec. 256, 165–176.

Muldoon, M.F., Manuck, S.B. and Matthews, K.A. (1990) Lowering cholesterol concentrations and mortality: a quantitative review of primary prevention trials. Br. Med. J. 301, 309–314.

Muldoon, M.F., Rossouw, J.E., Manuck, S.B., Glueck, C.J., Kaplan, J.R. and Kaufman, P.J. (1993) Low or lowered cholesterol and risk of death from suicide or trauma. Metabolism 42 (9 suppl.), 45–56.

Muldoon, M.F., Manuck, S.B., Mendelsohn, B. and Belle, J.R. Kaplan amd S.H. (2001) Cholesterol reduction and non-illness mortality: meta-analysis of randomised clinical trials. Br. Med. J. 322, 11–15.

New, A.S., Novotny, S.L. and Siever, M.S. Buchsbauma aand L.J. (1998) Neuroimaging in impulse-aggressive personality disorder patients, in: M. Maes and E.F. Coccaro (Eds.), Neurobiology and Clinical Views on Aggression and Impulsivity, Vol. 5, Wiley, New York, pp. 81–94.

New, A.S., Sevin, E.M., Mitropoulou, V., Reynolds, D., Novotny, S.L., Callahan, A., Trestman, R.L. and Siever, L.J. (1999) Serum cholesterol and impulsivity in personality disorders. Psychiatry Res. 85, 145–150.

Parthasarathy, S. and Rankin, S.M. (1992) Role of oxidized low density lipoprotein in atherogenesis. Progr. Lipid Res. 31, 127–143.

Pekkanen, L., Nissinen, A., Punsar, S. and Karvonen, M.J. (1989) Serum cholesterol and risk of accidental or violent death in a 25-year follow-up: the Finnish cohort of the seven countries study. Arch. Intern. Med. 149, 1589–1591.

Posse de Chaves, E.I., Rusinol, A.E., Vance, D.E., Campenpt, R.B. and Vance, J.E. (1997) Role of lipoproteins in the delivery of lipids to axons during axonal regeneration. J. Biol. Chem. 272, 30766–30773.

Raine, A., Brennan, P. and Mednick, S.A. (1994) Birth complications combined with early maternal rejection at age 1 year predispose to violent crime at age 18 years. Arch. Gen. Psychiatry 51, 984–988.

Raine, A., Meloy, J.R., Bihrle, S., Stoddard, J., LaCasse, L. and Buchsbaum, M.S. (1998) Reduced prefrontal and increased subcortical brain functioning assessed using positron emission tomography in predatory and affective murderers. Beh. Sci. Law 16, 319–332.

Sinclair, A.J., O'Dea, K., Dunstan, G., Ireland, P.D. and Niall, M. (1987) Effects on plasma lipids and fatty acid composition of very low fat diets enriched with fish or kangaroo meat. Lipids 22, 53–59.

Skinner, E.R. (1994) High density lipoprotein subclasses. Curr. Opin. Lipidol. Curr. Sci. 5, 241–247.

Skinner, E.R., Watt, C., Besson, J.A.O. and Best, P.V. (1993) Differences in the fatty acid composition of the grey and white matter of different regions of the brains of patients with Alzheimer's disease and control subjects. Brain 116, 717–725.

Skinner, F.K., Macdonnell, L.E.F., Glen, E.M.T. and Glen, A.I.M. (1989) Repeated automated assessment of abstinent male alcoholics: essential fatty acid supplementation and age effects. Alcohol Alcoholism 24, 129–139.

Stein, O., Stein, Y., Lefevre, M. and Roheim, P.S. (1986) The role of apolipoprotein A-IV in reverse cholesterol transport studied with cultured cells and liposomes derived from an ether analog of phosphatidylcholine. Biochim. Biophys. Acta 878, 7–13.

Stewart, M.A. and Stewart, S.G. (1981) Serum cholesterol in antisocial personality: a failure to replicate earlier findings. Neuropsychobiology 7, 9–11.

Sullivan, P.F., Joyce, P.R., Bulik, M., Mulder, R.T. and Oakley-Browne, M. (1994) Total cholesterol and suicidality in depression. Biol. Psychiatry 36, 472–477.

Terrisse, L., Poirier, J., Bertrand, A., Merrched, A., Visvikis, S., Siest, G., Milne, R. and Rassart, E. (1998) Increased levels of apolipoprotein D in cerebrospinal fluid and hippocampus of Alzheimer's patients. J. Neurochem. 71, 1643–1650.

Tilvis, R.S., Erkinjuntti, T., Sulkava, R. and Miettinen, T.A. (1987) Fatty acids of plasma lipids, red cells and platelets in Alzheimer's disease and vascular dementia. Atherosclerosis 65, 237–245.

Virkkunen, M. (1979) Serum cholesterol in antisocial personality disorder. Neuropsychobiology 5, 27–32.

Virkkunen, M. (1983) Serum cholesterol levels in homicidal offenders. A low cholesterol level is connected with a habitually violent tendency under the influence of alcohol. Neuropsychobiology 10, 65–69.

Virkkunen, M. and Penttinen, H. (1984) Serum cholesterol in aggressive conduct disorder: a preliminary study. Biol. Psychiatry 19, 435–439.

Virkkunen, M.E., Horrobin, D.F., Jenkins, D.K. and Manku, M.S. (1987) Plasma phospholipid essential fatty acids and prostaglandins in alcoholic, habitually violent, and impulsive offenders. Biol. Psychiatry 22, 1087–1096.

Weiger, W.A. and Bear, D.M. (1988) An approach to the neurology of aggression. J. Psychiatr. Res. 22, 85–98.

E.R. Skinner (Ed.), *Brain Lipids and Disorders in Biological Psychiatry*

CHAPTER 7

Do long-chain polyunsaturated fatty acids influence infant cognitive behavior?

J.S. Forsyth[1] and P. Willatts[2]

[1] *Department of Child Health, University of Dundee, Dundee, DD1 9SY, Scotland, UK;*
tel.; 01382 632974; fax: 01382 633921.
[2] *Department of Psychology, University of Dundee, Dundee, DD1 4HN, Scotland, UK;*
tel: 01382 344623; fax: 01382 229993

1. Introduction

Several observational studies have indicated that breast-fed children may be intellectually advantaged compared to children who were formula fed (Hoefer and Hardy 1929, Taylor 1977, Rodgers 1987, Lucas et al. 1992). The advantage persists after adjustments have been made for social, demographic, and educational factors that are known to influence developmental scores. Various constituents of human milk, including hormones, growth-promoting substances and nutritional components, have been proposed as potential factors that may positively influence neural development (Lucas et al. 1992). It has also been suggested, however, that the advantage to the infant may not be due to the breast milk per se but a consequence of the mother's decision to provide milk. A mother who is sufficiently concerned about her child's welfare to choose breast feeding, may also place greater emphasis on the social and developmental aspects of parental care (Pollock 1989).

In a large multi-center study undertaken by Lucas and colleagues (1992), preterm infants who received their mother's milk had a substantial advantage in IQ at age 7.5–8 years compared to those who did not receive their mother's milk. This study also showed a significant dose–response relation between the proportion of mother's milk consumed and later development. As the infants received their mother's milk by nasogastric tube, it was concluded that the observed advantage could not be due to interaction between mother and child during the process of suckling, and that factors present in breast milk are likely to influence cognitive ability. Moreover, as the infants only received their mother's milk for a period of 3–4 weeks, it was argued that the early postnatal period is a particularly sensitive period for brain development.

2. The developing brain

Following a full-term pregnancy, the average weight of an infant's brain is 350 grams. During the first year of life the brain weight increases by 750 g to 1.1 kg, with approximately 47% of this increase taking place in the cerebral cortex, and 60% of the weight gain being lipid, predominantly phospholipid (Cockburn 1997). During intrauterine life, most neurones have been formed by 22 weeks gestation and have migrated from their origin in the subependymal areas to the cortical surface. During the second half of pregnancy these cells develop complex arborizations, and after birth there is a marked increase in numbers of synaptosomes.

A major structural and functional component of synaptosomes is their phospholipid membrane. The principal membrane phospholipids are phosphatidylcholine (PC), phosphatidylethanolamine (PE), phosphatidylserine (PS) and phosphatidylinositol (PI). PC confers structural stability to the neural membrane (Cullis and De Krui 1979), while carboxyl groups of PS function as ion exchange sites (Cook et al. 1972). For nerve cells to fulfil their function of neurotransmission, the peptides and proteins responsible for enzymatic activity and neurotransmission in phospholipid membranes must be inserted into the membrane at the appropriate site and numbers. Control of the numbers and location of the peptides and proteins is not only dependent upon genetic factors but also upon the distribution and type of fatty acids attached to these phospholipids, in particular the chain length, degree of unsaturation and molecular configuration of the fatty acids. Docosahexaenoic acid (DHA) (C22:6 n-3) preferentially cross-links with proteins thus mediating activity of the membrane bound enzymes. The mechanisms that control the distribution of different phospholipids and their fatty acids are not known. However, deficiency of polyunsaturated fatty acids (PUFAs), including DHA, can change the physical properties and specific functions of the neural cell membrane, including enzyme action, receptor activity and signal transduction, and these in turn may affect the functioning level of the brain (Stubbs and Smith 1984).

Throughout pregnancy, essential fatty acids are required for growth and development of the placenta and fetus. It is estimated that about 600 grams of essential fatty acids are transferred from the mother to the fetus during a healthy term pregnancy. The fatty acid status of the mother is influenced by several factors including diet, parity and age. A woman's diet before conception plays an important role, as the mother's adipose tissue is a source of PUFA for the fetus during pregnancy. Fatty acid accretion by the fetal brain increases markedly towards the end of pregnancy, and averages 31.3 mg per week for total n-6 fatty acids and 14.5 mg per week for total n-3 fatty acids (Clandinin et al. 1989). Martinez (1991) estimated the average n-6 fatty acid accretion by the fetal forebrain to be 85 mg per week with arachidonic acid (C20:4 n-6) being the predominant long-chain PUFA and 30 mg per week for n-3 fatty acids, this being almost exclusively DHA.

3. Prenatal provision of LCPUFAs

3.1. Long-chain polyunsaturated fatty acids (LCPUFAs)

Long-chain polyunsaturated fatty acids are derived from the essential fatty acids, linoleic acid (18:2 n-6, LA) and alpha-linolenic acid (18:3 n-3, ALA). These parent fatty acids are substrates for lengthening and desaturation reactions which take place mainly in the liver endoplasmic reticulum, and which lead to the synthesis of LCPUFAs with 20–22 carbons and 2–6 double bonds. Desaturation enzymes are mixed-function oxygenases that are named according to the site at which the double bond is inserted. The position of desaturation is indicated by "delta" and occurs at carbon atoms 9, 6, 5, and 4 from the carboxyl group of the fatty acid. Recent studies have shown that delta-4 desaturation involves three enzymatic steps, rather than a single step by a delta-4 desaturation enzyme. The process begins with chain elongation, followed by delta-6 desaturation and then chain shortening by beta-oxidation in peroxisomes (Voss et al. 1991).

The rate-limiting step in fatty acid metabolism is desaturation, especially delta-6 desaturation. Fatty acids of the n-3, n-6, and n-9 series compete as substrates for the same desaturation enzymes. The delta-6 desaturation enzyme shows a preference for fatty acids in the order 18:3 n-3 > 18:2 n-6 > 18:1 n-9 (Brenner 1981). In addition, 25:5 n-3 and 24:4 n-6, the precursors for the delta-4 desaturation products will also compete for delta-6 desaturation. Desaturation of oleic acid (18:1 n-9) is more marked when 18:2 n-6 and 18:3 n-3 availability is low. A high level of Mead acid (20:3 n-9) is therefore considered an indicator of essential fatty acid deficiency.

Until recently, it was assumed that the endogenous synthesis of LCPUFAs would meet the requirements of the term and preterm infants. However, there is increasing evidence that, during the first months of life, the process of elongation and desaturation of the parent LCPUFAs linoleic acid and alpha-linolenic acid to their functional derivatives, arachidonic acid (AA) and docosahexaenoic acid (DHA) respectively, may be impaired in preterm and some term infants (Cook et al. 1972). As a result, there may be a deficiency of LCPUFAs at a critical time of brain growth and development.

3.2. Long-chain polyunsaturated fatty acids and neural function

Although it is recognized that LCPUFAs are major structural components of neural cell membranes, their precise functional role has still to be elucidated (Kurlak and Stephenson 1999). DHA is a major constituent of synaptic end sites while arachidonic acid is present in both the growth-cone region and the synaptosomes. During conversion of nerve growth cones to mature synapses, there is a substantial incorporation of DHA-containing lipids. The delivery of DHA to the growth cones is therefore likely to be a prerequisite for the formation of mature synapses (Martin and Bazan 1992). The effect of DHA on the structure of cellular membrane is to increase fluidity, and it has been shown that acetylcholine receptor function can be influenced by membrane fluidity. Arachidonic acid has been shown to be preferentially released from membrane phospholipids by the action of endogenous phospholipase A2 (Kim et al. 1996), and it is postulated that AA, as a free acid, has an important role in signal transduction (Kurlak and Stephenson 1999).

3.3. LCPUFA status of breast- and formula-fed infants

The exogenous supply of LCPUFAs to breast- and formula-fed infants has differed as LCPUFAs are present in breast milk but until recently have not been available in formula milks. Several studies have now reported that infants fed formulas that contain linoleic acid and alpha-linolenic acid but are devoid of LCPUFA, have lower plasma and erythrocyte membrane concentrations compared to breast-fed infants (Putman et al. 1982, Koletzko et al. 1989, Makrides et al. 1994, Decsi et al. 1995). More objective evidence is available from studies which have determined the relationship between brain fatty acids and diet in infancy and have demonstrated that breast-fed infants have higher concentrations of DHA in their cerebral cortex compared to infants fed formula milk (Farquharson et al. 1992, 1995, Makrides et al. 1994). Furthermore, the accumulation of DHA in the cerebral cortex was shown to be related to the length of breast feeding (Makrides et al. 1994). In contrast, there were no differences in the level of cortical arachidonic acid between breast- and formula-fed infants, suggesting that either AA is aggressively conserved in brain tissue or the amount of AA produced from the precursor in formula is sufficient to meet the AA requirements of the rapidly growing brain.

This evidence of relative deficiency of LCPUFAs in formula-fed infants gives support to the argument that artificial formulas should be supplemented with LCPUFAs. However, before recommendations can be made on LCPUFA supplementation, consideration needs to be given to the quantity and quality of potential supplements. The starting point for this work has been to detail the LCPUFA content of human milk.

3.4. LCPUFA content of human milk

Human milk contains between 6% and 14% of total fatty acids as linoleic acid and 0.6–2.0% of total fatty acids as alpha-linolenic acid. In addition, human milk also contains 0.3–1.0% AA and 0.1–0.9% DHA as well as other 20- and 22-carbon-atom n-6 and n-3 LCPUFAs (Jensen et al. 1995). Studies from Europe, Africa and Australia have demonstrated that levels of essential PUFAs linoleic and alpha-linolenic acid in human milk are influenced by the diet of lactating women but the concentrations of LCPUFAs are less affected (Koletzko et al. 1992, 1988, Gibson and Kneebone 1981). Moreover, the n-6 and n-3 LCPUFAs in milk were found not to be related to their respective precursors, linoleic and alpha-linolenic acids (Gibson and Kneebone 1984). In contrast, n-6 and n-3 LCPUFAs correlated with each other which is compatible with the hypothesis that n-6 and n-3 LCPUFAs share a mutual pathway for synthesis and/or secretion in milk. The average ratio of n-6 to n-3 was 2.7 for European milks and 2.2 for African milks. It has been speculated that because LCPUFAs are of greater biological relevance than the precursor fatty acids to the organism, the metabolic regulation of LCPUFAs serves to protect the breast-fed infant from wide variation in supply. However, it has been shown that if the dietary intake of preformed n-3 LCPUFA is significantly increased, this is reflected by an increase in n-3 LCPUFA concentrations in the milk (Innis et al. 1990, Carlson et al. 1986, Gibson and Kneebone 1981, Koletzko et al. 1988, Kneebone et al. 1985, Innis and Kuhnlein 1988). More recently a direct correlation was demonstrated between the level of DHA in breast milk and DHA in the maternal diet (Makrides et al. 1996). Clandinin and colleagues deduced that breast-fed infants receive a dietary LCPUFA supply which will meet their requirements for tissue accretion, even if they are born prematurely (Clandinin et al. 1989).

In human milk the LCPUFAs are predominantly esters of triacylglycerols. The rate of release of LCPUFAs from triacylglycerol by pancreatic colipase-dependent lipase at the sn-3 (and sn-1) positions is dependent upon the positions of double bonds, with a relative resistance to the release of AA and eicosapentaenoic acid (C20:5 n-3) from triacylglycerides compared to the more saturated fatty acids such as linoleic acid (C18:2 n-6) and oleic acid (C18:1 n-9). However, bile salt stimulated lipase (BSSL) efficiently hydrolyses the ester bonds of DHA and AA at all positions, even at the low levels of bile salt concentration which are present during the newborn period. These data not only show that an important function of BSSL in breast-fed infants is to enable optimal release and utilization of LCPUFAs from human milk, but also indicate that the metabolic fate of LCPUFA may differ between breast- and formula-fed infants (Chen et al. 1989, 1994, Hernell et al. 1993).

Moreover, there is evidence that the *sn*-position of DHA may determine the metabolic fate and tissue distribution of the fatty acid. In a study of newborn rats who were either fed oils with DHA located in the *sn*-2 position or DHA randomly distributed, the specifically structured triacyglycerol resulted in a higher level of DHA in the brain compared to the randomized oil which had higher levels of DHA in the liver (Christensen and Hey 1997). More recent data from this group indicate that these metabolic differences may have functional significance (Christensen et al. 1996).

3.5. Relation of essential precursors to LCPUFAs

The competitiveness between metabolic pathways for n-6 and n-3 fatty acid elongation and desaturation has important implications for recommendations on LCPUFA requirements for infants. In human milk the ratio of linoleic acid to alpha-linolenic acid is generally between 5 to 10. Ratios on either side of this range produced fatty acid profiles markedly different from that of breast-fed infants (Clark et al. 1992), and extremes of LA/ALA ratio altered 22-carbon fatty acid concentration in the brain (Dyer and Greenwood 1991). Previous recommendations on levels of AA and DHA in formulas have referred to the amounts of linoleic acid and alpha-linolenic acid in human milk and thus manufacturers have opted for an LA/ALA ratio of approximately 10:1. However, this does not allow for the preformed LCPUFAs in human milk (Gibson et al. 1994). The optimum LA/ALA ratio can therefore only be considered in parallel with the discussion on whether formulas should be supplemented with LCPUFA. It has been recommended, however, that the LA/ALA ratio should not be less than 5 and not greater than 15 (Commission of European Communities EC Directive 1996). In 1985 the American Academy of Pediatrics Committee on Nutrition recommended linoleic acid and alpha-linolenic acid intakes of at least 2.7% and 0.3% of energy, respectively. Functional studies published since these recommendations were made suggest that 0.3% energy from alpha-linolenic acid is probably too low even for term infants. However, excessively high concentrations of alpha-linolenic acid can be associated with growth impairment (Jensen et al. 1997).

3.6. LCPUFA supplement

The available evidence indicates that higher red cell membrane levels of AA and DHA are achieved if formulas are supplemented with preformed LCPUFA rather than the precursor essential fatty acids linoleic acid and alpha-linolenic acid. EC Directives (1996) have recently indicated that long-chain (C20 and C22) fatty acids may be added to infant formulas. However, the FDA has stated that further clinical studies evidencing functional benefit are required. For these studies to be undertaken, decisions on optimum quantity and quality of LCPUFA supplementation need to be clarified.

Attempts have been made to define the LCPUFA requirement of the term infant by estimating the intake of a breast-fed infant. The essential fatty acid intake of infants fed human milk has been calculated from the average composition of mature milk of European mothers (Koletzko et al. 1992) assuming an average fat content of 35 g/l and a daily intake of 175 ml/kg. The average supply of total LCPUFA is 100 g/kg/d (an amount that clearly exceeds intrauterine LCPUFA deposition in brain of approximately 6.5 to 16.5 mg/d (Martinez 1991, Clandinin et al. 1989). The average daily supply of linoleic acid was 660 mg/kg, alpha-linolenic acid 50 mg/kg, n-6 LCPUFA 70 mg/kg and n-3 LCPUFA 33 mg/kg. In human milk the n-6/n-3 LCPUFA ratio is kept relatively constant at about 2.0 (Koletzko et al. 1988), and brain lipids of newborn infants contain about twice as much n-6 LCPUFA as n-3 LCPUFA (Svennerholm 1968, Clandinin et al. 1989). A recommendation has therefore been made that n-6 LCPUFA content should not exceed 2% of total fat content and n-3 LCPUFA content should not exceed 1% (Statutory Instruments 1995).

The LCPUFA supplement needs to be safe and commercially viable. AA and DHA are not present in currently used vegetable oils, and alternative sources have therefore been explored. An early source of DHA was fish oil that also contains high levels of eicosapentaenoic acid (EPA, 20:5 n-3). There is very little EPA in human milk, and this preparation was associated with poorer growth in preterm infants (Carlson et al. 1992). More recent studies have tried to avoid the addition of EPA by using fish oils high in DHA, such as tuna oil (Auestad et al. 1997). Borage and evening primrose oil contain gamma-linolenic acid (GLA, 18:3 n-6) a metabolic precursor of AA that by-passes the rate-limiting step of delta-6 desaturase enzyme. Data from two studies indicate that GLA is not able to prevent a decline in AA levels of formula-fed infants (Makrides et al. 1995, Ghebrebeskel et al. 1995).

European studies have used egg phospholipids as a source of DHA and AA (Agostoni et al. 1995, Forsyth and Willatts 1996). Egg lipids have a high content of phosphatidylcholine and cholesterol (Agostoni et al. 1994), and since large quantities are required to achieve the LCPUFA levels in human milk, the amounts of lecithin and cholesterol could exceed that allowed as an additive (Ministry of Agriculture, Fisheries and Food 1992). In egg lipids the LCPUFAs are predominantly in phospholipids, whereas in human milk they are more evenly distributed between triglycerides and phospholipids. There are differences in absolute levels of LA and ALA, and the ratio of LA to ALA and proportions of saturates and mono-unsaturates can influence LCPUFA incorporation into cell membranes.

The production of LCPUFAs by micro-organisms in large-scale controlled conditions is a promising source. The two micro-organisms presently used for DHA and AA production are a common marine microalga *Crypthecodinium cohnii* and the soil fungus *Mortierella alpina* respectively. Since the sources of single-cell LCPUFAs have not previously been used as food for man, the oils are considered to be novel and consequently their safety and nutritional aspects are currently being assessed by the Advisory Committee on Novel Foods and Processes (Wells 1996).

4. LCPUFA and infant cognitive development

Improved cognitive abilities in children who were breast fed compared to children fed formula may be explained by breast-fed infants receiving higher levels of preformed dietary LCPUFA. However, studies of children who were breast fed or formula fed involve non-random assignment to these groups, and other factors such as genetic and social-demographic variables, parenting skills, or quality of parent–child interaction may also contribute to differences in cognitive abilities. Attempts to statistically control for the potentially confounding effects of these factors may be inadequate (Wright and Deary 1992), and several studies have reported that, although the cognitive advantage of breast feeding remained after adjustment for social and demographic variables, it disappeared after adjustment for parental IQ (Richards et al. 1998, Jacobson and Jacobson 1992, Jacobson et al. 1999). The role of LCPUFA in cognitive development must therefore be investigated by studies in which the effects of potentially confounding variables are controlled by randomizing infants to formulas containing either LCPUFA or no LCPUFA. Randomized studies of the effects of LCPUFA on development in both term and preterm infants have employed a variety of assessments which include tests of psychomotor development, language development, visual attention, and problem solving.

4.1. LCPUFA and infant psychomotor and language development

Standardized tests such as the Bayley Scales (Bayley 1993) provide global measures of infant psychomotor development. Carlson (1994) randomized preterm infants to formula containing either no LCPUFA or a supplement derived from fish oil containing both EPA and DHA. Bayley Mental Developmental Index (MDI) scores at corrected age 12 months did not differ between the groups, but Psychomotor Developmental Index (PDI) scores tended to be lower in the EPA + DHA supplemented group, a result that was consistent with the finding of poorer growth in preterm infants supplemented with DHA and relatively high levels of EPA (Carlson et al. 1992). In a second study in which the DHA content of the supplemented formula remained the same but the EPA content was reduced, Carlson (1994) reported significantly higher MDI scores in preterm infants aged 12 months who received n-3 LCPUFA in comparison to the control group. Damli et al. (1996) also reported higher MDI scores in 6-month-old preterm infants who were fed formula supplemented with AA + DHA derived from egg phospholipid in comparison to infants receiving no supplement.

Two randomized studies have examined the effects of LCPUFA supplementation on measures of psychomotor development in term infants. Agostoni et al. (1995) compared scores on the Brunet–Lézine test in two groups of infants fed a formula containing no LCPUFA or supplemented with AA + DHA derived from egg phospholipid. Significantly higher test scores were observed in the LCPUFA-supplemented group at age 4 months, but no differences were found in a follow-up conducted at age 24 months (Agostoni et al. 1997). In a multi-center study reported by Scott et al. (1997), scores on the Bayley Scales were compared in three groups of infants at age 12 months. The groups received either standard formula containing no LCPUFA, formula containing DHA from fish oil, or formula containing AA and DHA from egg phospholipid. No differences were observed in either MDI or PDI scores. Language development was also measured at age 14 months with the MacArthur Communicative Development Inventory (CDI). The CDI is a questionnaire completed by parents which provides information on infants' comprehension vocabulary, production vocabulary, and range of communicative gestures (Fenson et al. 1991). There were no differences between the groups on measures of comprehension vocabulary or communicative gestures, but the DHA-supplemented group had significantly lower scores on production vocabulary than the control group which received no LCPUFA supplementation. However, a follow-up study of the same children at age 3 years found no group differences on measures of IQ and vocabulary (Scott et al. 1997).

These results suggest that preformed dietary LCPUFA enhance the rate of infant psychomotor development during the first year of life, particularly in preterm infants who may have higher dietary n-3 requirements than term infants (Innis 1992). However, differences in test scores between diet groups may not reflect differences in cognitive abilities. Items on the Bayley Mental Scale for children younger than 2 years principally measure perceptual–motor abilities, and may not measure important cognitive changes that are occurring during this period (Roberts et al. 1999). Moreover, scores on standardized tests of infant development correlate poorly with measures of childhood intelligence (Slater 1995). For these reasons, researchers have employed other assessments to identify more specific effects of LCPUFA on infant cognitive function.

4.2. Infant visual information processing

Habituation is an inhibitory process involving decline in attention paid to a stimulus. Behaviorally, visual habituation is the decrease in looking time that occurs when a stimulus becomes familiar (Ruff and Rothbart 1996). The most frequently used technique for observing infant visual habituation is infant control procedure (ICP) in which the total looking time at a stimulus is determined after achievement of a criterion (Horowitz et al. 1972). The standard criterion is a reduction of at least 50% in the duration of infants' later looks compared to initial looks. Infant habituation involves active processing of information and memory for a stimulus. Recognition of the familiar (habituated) stimulus is demonstrated by longer looking at a novel stimulus compared to the familiar stimulus in a paired-comparison test involving presentation of both stimuli for a fixed duration (Bornstein 1985, Colombo 1993, Slater 1995). Relatively stable individual differences in look duration measures during habituation have been reported in the first year of life (Rose et al. 1986, Colombo et al. 1987). Look duration shows modest but significant correlations with childhood cognitive measures such as IQ and vocabulary scores (Colombo 1993, McCall and Carriger 1993, Slater 1995). These correlations are negative, with children who demonstrated more efficient habituation (shorter total looking times and shorter individual look durations) having higher IQ and vocabulary scores.

An alternative to habituation for investigating infant visual information processing and memory is the visual recognition memory (VRM) paradigm (Colombo 1993, Slater 1995). In the VRM paradigm, infants are initially familiarized to either a single stimulus or pair of identical stimuli for a fixed and relatively short duration. Infants subsequently receive a paired-comparison test in which the proportion of looking time directed at the novel stimulus (novelty-preference score) indicates the extent to which information about the familiarized stimulus was processed and encoded in memory. Infants who process more information during familiarization show greater preference for the novel stimulus in a paired-comparison test (Hunter et al. 1982, Rose et al. 1982, Richards 1997). An association has also been shown between lower novelty-preference scores and longer duration of individual looks on paired-comparison tests in low-birthweight infants (Rose et al. 1988), and in infants with prenatal exposure to alcohol (Jacobson et al. 1993) compared to normal controls. These results suggest that longer look durations in paired-comparison tests reflect less efficient information processing and poorer attention control. The Fagan Test of Infant Intelligence (Fagan and Shepard 1986) is based on the VRM paradigm. VRM novelty-preference correlates positively with measures of childhood IQ (Fagan and McGrath 1981, Rose et al. 1992), whereas look duration during paired-comparison tests correlates negatively with childhood IQ (Jacobson 1999).

4.3. Theoretical models of individual differences in fixation duration

Total looking times during habituation and novelty-preference scores in VRM tests may reflect differences in the speed at which information is processed and encoded into memory (Bornstein 1985, Colombo 1993). Infants who are faster at processing information have shorter looking times during habituation and higher novelty-preference scores in VRM tests. Evidence for such a relationship comes from studies in which infants were initially classified as short or long lookers depending on the duration of their longest (peak) fixation on a stimulus. In VRM tests, short lookers demonstrated a significant novelty preference after a relatively brief exposure to the familiar stimulus. In contrast, long lookers required a longer exposure to the familiar stimulus before showing a novelty preference (Colombo et al. 1991, Freeseman et al. 1993).

Other factors involving regulation of attention may also contribute to look duration. Differences in look duration may relate to infants' ability to disengage attention from a fixated stimulus when information processing is completed (Colombo 1995, McCall 1994). Young infants show considerable difficulty at releasing attention from one stimulus to orient to another, but substantial improvement in disengagement occurs after 3 months (Johnson et al. 1991, Hood 1995). Development of several cortical structures including parietal cortex, prefrontal cortex, and frontal eye fields is thought to be involved in this change (Hood 1995, Johnson 1994). Infants who have short look durations display better ability at disengaging attention from a stimulus than infants with long look durations (Frick et al. 1999). Ability at disengagement may also explain differences in VRM novelty-preference scores (McCall 1994). Infants with better ability at disengagement should quickly shift attention from the familiar to the novel stimulus, resulting in a higher novelty-preference score. Infants with poorer ability at disengagement should tend to look longer at the familiar stimulus before shifting fixation to the novel stimulus, resulting in a lower novelty-preference score.

Infant distractibility may also influence looking time. McCall (1994) proposed that infants begin processing stimulus information on their longest (peak) fixation, but that differences in distractibility, mediated by inhibitory processes, may affect the occurrence of the peak fixation. Infants with poorer inhibitory control may initially produce several short fixations on a stimulus because they are easily distracted by less salient stimuli, and their peak fixation would therefore occur late in the sequence of fixations. In contrast, infants with better inhibitory control would be less distractible and their peak fixation would occur when they first fixate the stimulus. Thus, infants who are more distractible may have more fixations and longer looking times. Support for this hypothesis comes from Lécuyer (1989) and Colombo et al. (1997) who reported that the peak fixation occurred earlier in the sequence of fixations in infants with shorter looking times.

Finally, look duration may be influenced by individual differences in arousal. Kaplan et al. (1991) proposed a dual-process model of infant look duration which is determined by the summation of two distinct processes – habituation and sensitization. Habituation involves a decrease in attention, whereas sensitization involves an increase in attention through activation of the arousal system, the degree of arousal being related to the amount of stimulus energy. Kaplan and Werner speculated that individual differences in susceptibility to the effects of stimulation on the arousal system may be related to the occurrence of early and late peak fixations and duration of looking. Infants who are easily aroused by stimulation (sensitizers) would tend to have later peak fixations and longer looking times than infants who are less easily aroused. However, there is currently little direct evidence to support this hypothesis (Colombo et al. 1997).

4.4. LCPUFA and infant visual information processing

Only one study has examined the relationship between LCPUFA supplementation and infant habituation (Forsyth and Willatts 1996). Term infants were randomized to formulas containing AA + DHA derived from egg phospholipid or no LCPUFA, and measures of visual habituation were obtained at age 3 months. Although LCPUFA-supplemented infants tended to have shorter fixation durations, none of the overall comparisons between the groups was significant. However, infants who had late peak fixations and who may have been more distractible had significantly shorter total fixation durations if they received LCPUFA supplementation in comparison to infants who received no LCPUFA. Infants with late peak fixations also had reduced growth parameters at birth compared to infants with early peak fixations. Duration of fixation in infants with early peak fixations was not influenced by LCPUFA supplementation. These results suggest that infants who have suffered moderate intra-uterine undernutrition may develop more efficient information processing if they are supplemented with LCPUFA.

Three randomized studies have examined the relationship between LCPUFA supplementation and performance on the Fagan Test of Infant Intelligence. Clausen et al. (1996) reported significantly higher novelty-preference scores in term infants fed a formula supplemented with AA + DHA from egg phospholipid compared to infants fed a formula containing no LCPUFA. In two studies of preterm infants (Carlson and Werkman 1996, Werkman and Carlson 1996), infants received formula containing either no LCPUFA or DHA + EPA from fish oil which had a relatively high ratio of DHA to EPA. No significant differences between the diet groups were found on novelty-preference scores, but fixation durations in paired-comparison tests were shorter in the supplemented groups. In the study by Carlson and Werkman (1996), LCPUFA supplementation was stopped at age 2 months but the effects on look duration were seen at age 12 months. Similar results were obtained by Reisbick et al. (1997) in a study of VRM in infant rhesus monkeys fed diets containing adequate or low levels of alpha-linolenic acid. There was no effect of diet group on novelty preference, but infant monkeys who received an adequate level of alpha-linolenic acid had shorter fixation durations than monkeys who received a low level of alpha-linolenic acid.

The consistency of these findings suggests that preformed dietary LCPUFA (and in particular, n-3 LCPUFA) are important for more efficient infant visual information processing, a conclusion that is supported by the findings with non-human primates. These effects do not appear to be related to the reported effects of LCPUFA on visual acuity development (e.g., Makrides et al. 1995, Birch et al. 1998) because look duration and acuity scores are not correlated (Carlson et al. 1995, Neuringer et al. 1996, Jacobson 1999). There is some evidence that the effects on visual attention persist beyond the period that infants receive supplementation (Carlson and Werkman 1996). However, it is not possible to determine whether LCPUFA influence speed of processing, ability at disengagement, or other processes involved in attention regulation because the assessments employed in these studies cannot distinguish between these alternative factors. The question of which processes are influenced by LCPUFA requires further study with more specific tests of visual attention that are appropriate for detecting differences in both speed of processing and attention control.

4.5. Infant means–end problem solving

Means–end problem solving involves the deliberate and planful execution of a sequence of steps to achieve a goal, and this ability develops rapidly after 6 months of age (Willatts 1989). Infants aged between 7 and 8 months begin to solve simple one-step problems such as searching under a cover for a toy, or pulling a cloth to retrieve a toy that is resting on it (Willatts 1984, 1999). At 9 months, infants can solve more complex problems requiring the completion of two intermediate steps to achieve a goal (Willatts 1997). Two-step problem-solving scores measured at 9 months correlate positively with IQ and vocabulary scores measured at 3 years (Slater 1995, Willatts 1997). Infants at 10 months can solve more complex problems which require the execution of three intermediate steps to achieve a goal (Willatts and Rosie 1992). It is not known whether ability at three-step problem solving is related to childhood cognitive scores.

Means–end problem solving involves planning, sequencing actions, and maintaining attention to a goal, all of which are mediated by prefrontal cortex (Diamond 1991, Johnson 1997, Roberts et al. 1998). There is some evidence linking infants' ability at means–end problem solving to development of prefrontal cortex (Bell and Fox 1992, Diamond et al. 1997), but a direct link between performance on multi-step means–end problems and prefrontal cortex has yet to be established.

4.6. LCPUFA and infant problem solving

Willatts et al. (1998a) examined the relationship between LCPUFA supplementation and problem solving at age 9 months in term infants. The infants in this study were randomized to formulas containing AA + DHA or no LCPUFA for a period of 4 months, and had previously been assessed on a test of visual habituation at age 3 months (Forsyth and Willatts 1996). Infants received several trials on a two-step problem in which a toy was placed on the end of a cloth and hidden under a cover. To solve the problem of retrieving the toy, infants first pulled the cloth to retrieve the cover, and then removed the cover to find the toy. Infants' sequence of behaviors was scored according to certain criteria for evidence of intention to retrieve the hidden toy. Intentional behavior was defined as behavior aimed at achieving the goal of retrieving the toy, rather than simply playing with the cloth or cover (Willatts 1984, 1999). There were two measures of problem-solving ability. Intention score was the total of the scores for all behaviors, averaged across trials, with a higher score indicating more intentional behavior. Number of intentional solutions was the number of trials on which infants solved the problem and demonstrated evidence of intention on all behaviors. LCPUFA-supplemented infants tended to have higher problem-solving scores than infants who received no LCPUFA, but the differences were not significant. However, among the infants who showed poorer attention control at 3 months (indicated by a late peak fixation), those who received the supplemented formula produced more intentional solutions than infants who received no LCPUFA.

The same infants were tested on a more complex problem at age 10 months (Willatts et al. 1998b). Infants had to complete three intermediate steps to achieve a final goal and solve the problem (remove a barrier to grasp a cloth, pull the cloth to retrieve a cover, and search under the cover to find a hidden toy). Infants who received the LCPUFA-supplemented formula had significantly higher problem-solving scores than infants who received no LCPUFA.

These results suggest that problem solving is improved in term infants who received formula supplemented with LCPUFA, and the benefits of LCPUFA supplementation to infants persist beyond the period that they received their supplemented formula. The differences between the diet groups may reflect faster information processing and therefore more efficient problem solving in the LCPUFA-supplemented group. Alternatively, they may reflect improved ability at disengagement in the LCPUFA-supplemented group. Infants who can easily disengage attention from a stimulus may be able to switch rapidly from manipulating one intermediary to the next, and so solve a complex means–end problem.

5. Conclusions

Although the number of studies of the effects of preformed dietary LCPUFA on infant cognitive function is relatively small, a consistent pattern of findings is beginning to emerge. In the majority of studies, both term and preterm infants fed artificial formula supplemented with LCPUFA (DHA or AA + DHA) showed improved cognitive performance in comparison to control infants fed a formula containing no LCPUFA. No negative effects on development have been observed with formulas containing both AA and DHA. Only two studies have reported poorer cognitive scores in groups receiving LCPUFA supplement (Carlson 1994, Scott et al. 1997). These studies share two features that may be related to these effects. First, the supplement was DHA without AA. Second, the DHA source was marine oil that also included EPA. This suggests that the ratios of EPA to DHA and AA to DHA are more relevant to cognitive outcome than absolute level of DHA. This conclusion is reinforced by the fact that in studies reporting a significant cognitive advantage in infants who received LCPUFA supplement, formulas were supplemented with AA + DHA or DHA with low EPA.

How can we explain the effects of LCPUFA on infant cognitive function? Studies showing improved scores on tests of psychomotor development in supplemented infants provide no information about which cognitive functions may have been influenced. Studies of visual attention and problem solving are more informative because the measures that are influenced by LCPUFA predict childhood cognitive scores. These measures therefore index relatively stable, central cognitive processes. Whether LCPUFA influence speed of processing, disengagement, or other processes involved in attention regulation remains uncertain. However, new assessments for examining these different processes have been evaluated and may offer methods for identifying the specific effects of LCPUFA on infant cognitive development. Moreover, evidence linking performance on assessments to different neurological systems (Diamond et al. 1997, Johnson 1997) may provide information on brain mechanisms.

Little is known about the long-term effects of early dietary LCPUFA on later cognitive development. In the majority of randomized studies, assessments have been undertaken during the first year of life, although some have reported effects that were observed beyond the period that infants received their LCPUFA supplement. Only two studies have included children older than one year, but neither found any differences between the diet groups on tests of psychomotor development, language, or IQ (Agostoni et al. 1997, Scott et al. 1998). It is possible that global tests of psychomotor development and IQ may not be sufficiently sensitive to the effects of LCPUFA on cognition because test scores reflect a broad range of abilities. Studies of human infants and nonhuman primates suggest that speed of processing and attention regulation may be influenced by LCPUFA, and it may be more appropriate to focus on tests of these abilities in future investigations of the long-term effects of LCPUFA on cognition.

References

Agostoni, C., Riva, E., Bellù, R., Trojan, S., Luotti, D. and Giovannini, M. (1994) Effects of diet on the lipid and fatty acid status of full term infants at 4 months. J. Am. Coll. Nutr. 13, 658–664.

Agostoni, C., Trojan, S., Bellù, R., Riva, E. and Giovannini, M. (1995) Neurodevelopmental quotient of healthy term infants at 4 months and feeding practice: The role of long chain polyunsaturated fatty acids. Pediatr. Res. 38, 262–266.

Agostoni, C., Trojan, S., Bellù, R., Riva, E., Bruzzese, M.G. and Giovannini, M. (1997) Developmental quotient at 24 months and fatty acid composition of diet in early infancy: A follow up study. Archiv. Dis. Child. 76, 421–424.

Auestad, N., Michael, B., Montalto, R.T.H., Hall, R.T., Fitzgerald, K.M., Wheeler, R.E., Connor, W.E., Neuringer, M., Connor, S.L., Taylor, J.A. and Hartmann, E.E. (1997) Visual acuity, erythrocyte fatty acid composition, and growth in term infants fed formulas with long chain polyunsaturated fatty acids for one year. Pediatr. Res. 41, 1–10.

Bayley, N. (1993) The Bayley Scales of Infant Development, 2nd edition, The Psychological Corporation, San Antonio, TX.

Bell, M.A. and Fox, N. (1992) The relations between frontal brain electrical activity and cognitive development during infancy. Child Dev. 63, 1142–1163.

Birch, E.B., Hoffman, D.R., Uauy, R., Birch, D.G. and Prestidge, C. (1998) Visual acuity and the essentiality of docosahexaenoic acid and arachidonic acid in the diet of term infants. Pediatr. Res. 44, 201–209.

Bornstein, M.H. (1985) Habituation of attention as a measure of visual information processing in human infants: Summary, systematization, and synthesis. In: G. Gottlieb and N.A. Krasnegor (Eds.), Measurement of audition and vision in the first year of postnatal life: A methodological overview, Ablex, Norwood, pp. 253–300.

Brenner, R.R. (1981) Nutritional and hormonal factors influencing desaturation of essential fatty acids. Prog. Lipid Res. 20, 41–47.

Carlson, S.E. (1994) Growth and development of premature infants in relation to ω-3 and ω-6 fatty status. In: C. Gali, A.P. Simopolous and E. Tremoli (Eds.), Fatty Acids and Lipids: Biological aspects, Karger, Basel, pp. 63–69.

Carlson, S.E. and Werkman, S.H. (1996) A randomized trial of visual attention of preterm infants fed docosa-hexaenoic acid until two months. Lipids 31, 85–90.

Carlson, S.E., Rhodes, P.G. and Ferguson, M.G. (1986) Docosahexaenoic acid and status of preterm infants at birth and following feeding with human milk or formula. Am. J. Clin. Nutr. 44, 798–804.

Carlson, S.E., Rhodes, P.G. and Ferguson, M.G. (1992) First year growth of preterm infants fed standard compared to marine oil n-3 supplemented formula. Lipids 27, 901–907.

Carlson, S.E., Werkman, S.H. and Peeples, J.M. (1995) Early visual acuity does not correlate with later evidence of visual processing in preterm infants although each is improved by the addition of docosahexaenoic acid to infants formula. In: Abstracts Second International Congress of the International Society for the Study of Fatty Acids (ISSFAL), American Oil Chemist's Society, Champaign, IL, p. 53.

Chen, Q., Sternby, B. and Nilsson, A. (1989) Hydrolysis of triacylglycerol arachidonic acid and linoleic acid ester bonds by human pancreatic lipase and carboxylester lipase. Biochim. Biophys. Acta 1004, 372–386.

Chen, Q., Blackberg, L., Nilsson, A., Sternby, B. and Hernell, O. (1994) Digestion of triacylglycerols containing long chain polyenoic fatty acids in vitro by colipase-dependent pancreatic lipase and human milk bile salt stimulated lipase. Biochim. Biophys. Acta. 1210, 239–243.

Christensen, M.M. and Hey, C.-E. (1997) Early dietary intervention with structured triacylglycerols containing docosahexaenoic acid. Effect on brain, liver, and adipose tissue lipids. Lipids 32, 185–191.

Christensen, M.M., Lund, S.P., Simonsen, L., Hass, U., Simonsen, E. and Hoy, C.-K. (1996) Visual and hearing performance and learning ability of rats given structured triacylglycerols containing docosahexaenoic acid from birth. Correlation with tissue fatty acid composition. In: Abstracts AOCS Conference, PUFA in Infant Nutrition: Consensus and Controversies, American Oil Chemist's Society, Champaign, IL, p. 23.

Clandinin, M.T., Chappel, J.E. and van Aerde, J.E.E. (1989) Requirements of newborn infants for long chain polyunsaturated fatty acids. Acta Paediatr. Scand. 78(Suppl. 351), 63–71.

Clark, K.J., Makrides, M., Neumann, M.A. and Gibson, R.A. (1992) Determination of the optimal ratio of linoleic acid to alpha-linolenic acid in infant formulas. J. Pediatr. 120, S151–S158.

Clausen, U., Damli, A., von Schenck, U. and Koletzko, B. (1996) Influence of long-chain polyunsaturated fatty acids (LCPUFA) on early visual acuity and mental development of term infants. In: Abstracts AOCS Conference, PUFA in Infant Nutrition: Consensus and Controversies, American Oil Chemist's Society, Champaign, IL, p. 12.

Cockburn, F. (1997) Fatty acids in early development. Prenat. Neonat. Med. 2, 108–115.

Colombo, J. (1993) Infant Cognition: Predicting later Intellectual Functioning. Sage, London.

Colombo, J. (1995) On the neural mechanisms underlying developmental and individual differences in visual fixation in infancy: Two hypotheses. Dev. Rev. 15, 97–135.

Colombo, J., Mitchell, D.W., O'Brien, M. and Horowitz, F.D. (1987) The stability of visual habituation during the first year of life. Child Dev. 58, 474–487.

Colombo, J., Mitchell, D.W., Coldren, J.T. and Freeseman, L.J. (1991) Individual differences in infant visual attention: Are short lookers feature processors or faster processors? Child Dev. 62, 1247–1257.

Colombo, J., Frick, J.E. and Gorman, S.A. (1997) Sensitization during visual habituation sequences: procedural effects and individual differences. J. Exp. Child Psychol. 67, 223–235.

Commission of the European Communities (1996) Commission Directive of 16 Feb 1996 (amendment of Directive 91/32/EEC) on infant formulae and follow-on formulae, Official J. Eur. Communities L 49, 12–16.

Cook, A.M., Low, E. and Ishijimi, M. (1972) Effect of phosphatidylserine decarboxylase on neuronal excitation. Nature New Biol. 239, 150–151.

Cullis, P.R. and De Krui, J.F.F.B. (1979) Lipid polymorphism and the functional roles of lipids in biological membranes. Biochim. Biophys. Acta 559, 339–420.

Damli, A., von Schenck, U., Clausen, U. and Koletzko, B. (1996) Effects of long-chain polyunsaturated fatty acids (LCPUFA) on early visual acuity and mental development in preterm infants. In: Abstracts AOCS Conference, PUFA in Infant Nutrition: Consensus and Controversies, American Oil Chemist's Society, Champaign, IL, p. 14.

Decsi, T., Thiel, I. and Koletzko, B. (1995) Essential fatty acids in full term infants fed breast milk or formula. Arch. Dis. Child. 72, F23–F28.

Diamond, A. (1991) Neuropsychological insights into the meaning of object concept development. In: S. Carey and R. Gelman (Eds.), The Epigenesis of Mind: Essays on Biology and Cognition, Erlbaum, Hillsdale, pp. 67–110.

Diamond, A., Prevor, M.B., Callender, G. and Druin, D.P. (1997) Prefrontal cortex deficits in children treated early and continuously for PKU. Monogr. Soc. Res. Child Dev. Ser. No. 252, 62(4).

Dyer, J.R. and Greenwood, C.E. (1991) Neural 22-carbon fatty acids in the weanling rat respond rapidly and specifically to a range of dietary linoleic to alpha-linolenic fatty acid ratios. J. Neurochem. 56, 1921–1931.

Fagan, J.F. and McGrath, S. (1981) Infant recognition memory and later intelligence. Intelligence 5, 121–130.

Fagan, J.F. and Shepard, P. (1986) The Fagan Test of Infant Intelligence, InfanTest Corporation, Cleveland, OH.

Farquharson, J., Cockburn, F., Patrick, W.A. and Jamieson, E.C. and Logan, R.W. (1992) Infant cerebral cortex phospholipid fatty acid composition and diet. Lancet 340, 810–813.

Farquharson, J., Jamieson, E.C., Abbasi, K.A., Patrick, W.J.A., Logan, R.W. and Cockburn, F. (1995) Effect of diet on the fatty acid composition of the major phospholipids of infant cerebral cortex. Arch. Dis. Child. 72, 198–203.

Fenson, L., Dale, P.S., Reznick, J.S., Thal, D., Bates, E., Hartung, J.P., Pethick, S. and Reilly, J.S. (1991) Technical manual for the MacArthur Communicative Development Inventories, San Diego State University, San Diego, CA.

Forsyth, J.S. and Willatts, P. (1996) Do LCPUFA influence infant cognitive behavior? In: J.G. Bindels, A.C. Goedhart and H.K.A. Visser (Eds.), Recent Developments in Infant Nutrition, Kluwer Academic, Dordrecht, pp. 225–234.

Freeseman, L.J., Colombo, J. and Coldren, J.T. (1993) Individual differences in infant visual attention: Four-month-olds' discrimination and generalization of global and local stimulus properties. Child Dev. 64, 1191–1203.

Frick, J.E., Colombo, J. and Saxon, T.F. (1999) Individual and developmental differences in disengagement of fixation in early infancy. Child Dev. 70, 537–548.

Ghebrebeskel, K., Leighfield, M. and Leaf, A. (1995) Fatty acid composition of plasma and red cell phospholipids of preterm infants fed on breast milk or formula. Eur. J. Pediatr. 154, 46–52.

Gibson, R.A. and Kneebone, G.M. (1981) Fatty acid composition of human colostrum and mature breast milk. Am. J. Clin. Nutr. 34, 252–257.

Gibson, R.A. and Kneebone, G.M. (1984) A lack of correlation between linoleate and arachidonate in human breast milk. Lipids 19, 469–471.

Gibson, R.A., Makrides, M., Neumann, M., Simmer, K., Mantzioris, E. and Jame, M.J. (1994) Ratios of linoleic acid to alpha-linolenic acid in formulas for term infants. J. Pediatr. 125, S48–S55.

Hernell, O., Blackberg, L., Chen, Q., Sternby, B. and Nilsson, A. (1993) Does the bile salt-stimulated lipase of human milk have a role in the use of the milk long chain polyunsaturated fatty acids? J. Pediatr. Gastroenterol. Nutr. 16, 426–431.

Hoefer, A. and Hardy, M.C. (1929) Later development of breast fed and artificially fed infants. J. Am. Med. Assoc. 92, 615–619.

Hood, B.M. (1995) Shifts of visual attention in the human infant: A neuroscientific approach. In: C. Rovee-Collier and L.P. Lipsitt (Eds.), Advances in Infancy Research, Vol. 9, Ablex, Norwood, pp. 163–216.

Horowitz, F.D., Paden, L.Y., Bhana, K. and Self, P.A. (1972) An infant control procedure for the study of infant visual fixations. Dev. Psych. 7, 90.

Hunter, M.A., Ross, H.S. and Ames, E.W. (1982) Preferences for familiar or novel toys: Effects of familiarization time in 1-year-olds. Dev. Psychol. 18, 519–529.

Innis, S.M. (1992) N-3 fatty acid requirements of the newborn. Lipids 27, 879–885.

Innis, S.M. and Kuhnlein, H.V. (1988) Long chain n-3 fatty acids in breast milk of Inuit women consuming traditional foods. Early Hum. Dev. 18, 185–189.

Innis, S.M., Foote, K.D., MacKinnon, M.J. and King, D.J. (1990) Plasma and red cell fatty acids of low birth weight infants fed their mother's expressed breast milk or preterm infant formula. Am. J. Clin. Nutr. 51, 994–1000.

Jacobson, S.W. (1999) Assessment of long-chain polyunsaturated fatty acid nutritional supplementation on infant neurobehavioral development and visual acuity. Lipids 34, 151–160.

Jacobson, S.W. and Jacobson, J.L. (1992) Breastfeeding and intelligence. Lancet 339, 926.

Jacobson, S.W., Jacobson, J.L., Sokol, R.J., Martier, S.S. and Ager, J.W. (1993) Prenatal alcohol exposure and infant information processing ability. Child Dev. 64, 1706–1721.

Jacobson, S.W., Chiodo, L. and Jacobson, J.J. (1999) Breastfeeding effects on intelligence quotient in 4- and 11-year-old children. Pediatrics 103, e70.

Jensen, C.L., Prager, T.C., Fraley, J.K., Chen, H., Anderson, R.E. and Heird, W.C. (1997) Effect of dietary linoleic/alpha-linolenic acid ratio on growth and visual function of term infants. J. Pediatr. 131, 200–209.

Jensen, R.G., Bitman, J., Carlson, S.E., Cavel, S.C., Hamosh, M. and Newburg, D.S. (1995) Milk lipids. A. Human milk lipids. In: R.G. Jensen (Ed.), Handbook of Milk Composition, Academic Press, San Diego, CA, pp. 495–542.

Johnson, M.H. (1994) Dissociating components of visual attention: A neurodevelopmental approach. In: M.J. Farah and G. Ratcliff (Eds.), The Neuropsychology of High-Level Vision, Erlbaum, Hillsdale, pp. 735–747.

Johnson, M.H. (1997) Developmental Cognitive Neuroscience, Blackwell, Oxford.

Johnson, M.H., Posner, M.I. and Rothbart, M.K. (1991) Components of visual orienting in early infancy: Contingency learning, anticipatory looking, and disengaging. J. Cogn. Neurosci. 3, 335–343.

Kaplan, P.S., Werner, J.S. and Rudy, J.W. (1991) Implications of a sensitization process for the analysis of infant visual attention. In: M.J.S. Weiss and P.R. Zelazo (Eds.), Newborn Attention: Biological Constraints and the Influence of Experience, Ablex, Norwood, pp. 278–307.

Kim, H.Y., Edsall, L. and Ma, Y.C. (1996) Specificity of polyunsaturated fatty acid release from rat brain synaptosomes. Lipids 31, S229–S333.

Kneebone, G.M., Kneebone, R. and Gibson, R.A. (1985) Fatty acid composition of breast milk from three racial groups from Penang, Malaysia. Am. J. Clin. Nutr. 41, 765–769.

Koletzko, B., Mrotzek, M. and Bremer, H.J. (1988) Fatty acid composition of mature milk in Germany. Am. J. Clin. Nutr. 47, 954–957.

Koletzko, B., Schmidt, E., Bremer, H.J., Haug, M. and Harzer, G. (1989) Effects of dietary long-chain polyunsaturated fatty acids on the essential fatty acid status of premature infants. Eur. J. Pediatr. 148, 669–675.

Koletzko, B., Thiel, I. and Abiodun, P.O. (1992) Fatty acid composition of human milk in Europe and Africa. J. Pediatr. 120, 562–570.

Kurlak, L.O. and Stephenson, T.J. (1999) Plausible explanations for effects of long chain polyunsaturated fatty acids (LCPUFA) on neonates. Arch. Dis. Child. 80, F148–F154.

Lécuyer, R. (1989) Habituation and attention, novelty and cognition: Where is the continuity? Hum. Dev. 32, 148–157.

Lucas, A., Morley, R., Cole, T.J., Lister, G. and Leeson-Payne, C. (1992) Breast milk and subsequent intelligence quotient in children born preterm. Lancet 339, 261–264.

Makrides, M., Neumann, M.A., Byard, R.W., Simmer, K. and Gibson, R.A. (1994) Fatty acid composition of brain, retina, and erythrocytes in breast and formula-fed infants. Am. J. Clin. Nutr. 60, 189–194.

Makrides, M., Neumann, M., Simmer, K., Pater, J. and Gibson, R.A. (1995) Are long-chain polyunsaturated fatty acids essential nutrients in infancy? Lancet 345, 1463–1468.

Makrides, M., Neumann, M.A. and Gibson, R.A. (1996) Effect of maternal docosahexaenoic acid (DHA) supplementation on breast milk composition. Eur. J. Clin. Nutr. 50, 352–357.

Martin, R.E. and Bazan, N.G. (1992) Changing fatty acid content of growth cone lipids prior to synaptogenesis. J. Neurochem. 59, 318–325.

Martinez, M. (1991) Developmental profiles of polyunsaturated fatty acids in the brain of normal infants and patients with peroxisomal diseases: Severe deficiency of docosahexaenoic acid in Zellweger's and pseudo-Zellweger's syndromes. World Rev. Nutr. Diet 61, 87–96.

McCall, R.B. (1994) What process mediates predictions of childhood IQ from infant habituation and recognition memory? Speculations on the role of inhibition and rate of information processing. Intelligence 18, 107–125.

McCall, R.B. and Carriger, M.S. (1993) A meta-analysis of infant habituation and recognition memory performance as predictors of later IQ. Child Dev. 64, 57–79.

Ministry of Agriculture, Fisheries and Food (1992) Food Advisory Committee Report on the use of additives in foods specially prepared for infants and young children, HMSO, London.

Neuringer, M., Reisbick, S. and Teemer, C. (1996) Relationships between visual acuity and visual attention measures in rhesus monkey infants. Invest. Ophthalmol. Vis. Sci. 37, S532.

Pollock, J.I. (1989) Mother's choice to provide breast milk and developmental outcome. Arch. Dis. Child. 89, 763–764.

Putman, J.C., Carlson, S.E., de Voe, P. and Barness, L.A. (1982) The effect of variations of dietary fatty acid composition on erythrocyte phosphatidylethanolamine in human infants. Am. J. Clin. Nutr. 36, 106–114.

Reisbick, S., Neuringer, M., Gohl, E., Wild, R. and Anderson, G.J. (1997) Visual attention in infant monkeys: Effects of dietary fatty acids and age. Dev. Psychol. 33, 387–395.

Richards, J.E. (1997) Effects of attention on infants' preference for briefly exposed stimuli in the paired-comparison recognition–memory paradigm. Dev. Psychol. 33, 22–31.

Richards, M., Wadsworth, M., Rahimi-Foroushani, A., Hardy, R., Kuh, D. and Paul, A. (1998) Infant nutrition and cognitive development in the first offspring of a national UK birth cohort. Dev. Med. Child Neurol. 40, 163–167.

Roberts, A.C., Robbins, T.W. and Weiskrantz, L. (1998) The Prefrontal Cortex: Executive and Cognitive Functions, Oxford University Press, Oxford.

Roberts, E., Bornstein, M.H., Slater, A. and Barrett, J. (1999) Early cognitive development and parental education. Infant Child Dev. 8, 49–62.

Rodgers, B. (1987) Feeding in infancy and later ability and attainment: a longitudinal study. Dev. Med. Child Neurol. 20, 421–426.

Rose, D., Slater, A. and Perry, H. (1986) Prediction of childhood intelligence from habituation in early infancy. Intelligence 10, 251–263.

Rose, S.A., Gottfried, A.W., Melloy-Carminar, P. and Bridger, W.H. (1982) Familiarity and novelty preferences in infant recognition memory: Implications for information processing. Dev. Psychol. 18, 705–713.

Rose, S.A., Feldman, J. and McCarton, C.M. (1988) Information processing in seven-month-old infants as a function of risk status. Child Dev. 59, 589–603.

Rose, S.A., Feldman, J. and Wallace, I. (1992) Infant information processing in relation to six-year cognitive outcomes. Child Dev. 63, 1126–1141.

Ruff, H.A. and Rothbart, M.K. (1996) Attention in Early Development: Themes and Variations, Oxford University Press, Oxford.

Scott, D.T., Janowsky, J.S., Hall, R.T., Wheeler, R.E., Jacobsen, C.H., Auestad, N. and Montalto, M.B. (1997) Cognitive and language assessment of 3.25 yr old children fed formula with or without long chain polyunsaturated fatty acids (LCP) in the first year of life. Pediatr. Res. 41, 240A.

Scott, D.T., Janowsky, J.S., Carroll, R.E., Taylor, J.A., Auestad, N. and Montalto, M.B. (1998) Formula supplementation with long-chain polyunsaturated fatty acids: Are there developmental benefits? Pediatrics. 102, E59.

Slater, A. (1995) Individual differences in infancy and later IQ. J. Child Psychol. Psychiatry 36, 69–112.

Statutory Instruments (1995) The Infant Formula and Follow-on Formula Regulations, HMSO, London.

Stubbs, C.D. and Smith, A.D. (1984) The modification of mammalian membrane polyunsaturated fatty acid composition in relation to membrane fluidity and function. Biochim. Biophys. Acta 779, 89–137.

Svennerholm, L. (1968) Distribution and fatty acid composition of phosphoglycerides in normal human brain. J. Lipid Res. 9, 570–579.

Taylor, B. (1977) Breast versus bottle feeding. N.Z. Med. J. 85, 235–238.

Voss, A., Reinhart, M., Sankarappa, S. and Sprecher, H. (1991) The metabolism of 7,10,13,16,19-docosa-pentaenoic acid to 4,7,10,13,16,19-docosahexaenoic acid in rat liver is independent of a 4-desaturase. J. Biol. Chem. 20, 41–47.

Wells, J.C.K. (1996) Nutritional considerations in infant formula design. Semin. Neonatol. 1, 19–26.

Werkman, S.H. and Carlson, S.E. (1996) A randomized trial of visual attention of preterm infants fed docosa-hexaenoic acid until nine months. Lipids 31, 91–97.

Willatts, P. (1984) Stages in the development of intentional search by young infants. Dev. Psychol. 20, 389–396.

Willatts, P. (1989) Development of problem solving in infancy. In: A. Slater and J.G. Bremner (Eds.), Infant Development, Erlbaum, London, pp. 143–182.

Willatts, P. (1997) Beyond the 'Couch Potato' infant: How infants use their knowledge to regulate action, solve problems, and achieve goals. In: J.G. Bremner, A. Slater and G. Butterworth (Eds.), Infant Development: Recent Advances, Psychology Press, Hove, pp. 109–135.

Willatts, P. (1999) Development of means–end behavior in young infants: Pulling a support to retrieve a distant object. Dev. Psychol. 35, 651–667.

Willatts, P. and Rosie, K. (1992) Thinking ahead: Development of means–end planning in young infants. Infant Beh. Dev. 15, 769.

Willatts, P., Forsyth, J.S., DiModugno, M.K., Varma, S. and Colvin, M. (1998a) Influence of long-chain polyunsaturated fatty acids on infant cognitive function. Lipids 33, 973–980.

Willatts, P., Forsyth, J.S., DiModugno, M.K., Varma, S. and Colvin, M. (1998b) Effect of long-chain polyunsaturated fatty acids in infant formula on problem solving at 10 months of age. Lancet 352, 688–691.

Wright, P. and Deary, I.J. (1992) Breastfeeding and intelligence. Lancet 339, 612.

E.R. Skinner (Ed.), *Brain Lipids and Disorders in Biological Psychiatry*

CHAPTER 8

Molecular species of phospholipids during brain development

Their occurrence, separation, and roles

Akhlaq A. Farooqui, Tahira Farooqui and Lloyd A. Horrocks

Department of Medical Biochemistry, The Ohio State University, Columbus, Ohio 43210, USA

1. Introduction

Glycerophospholipids are amphipathic molecules that form the backbone of bio-membranes where they are organized in bilayers and held together by hydrophobic, coulombic and Van der Waal forces and hydrogen bonds. Brain tissue contains four major classes of glycerophospholipids. The first three, 1,2-diacyl glycerophospholipids, 1-alk-1'-enyl-2-acyl glycerophospholipids or plasmalogens, and 1-alkyl-2-acyl glycero-phospholipids, have a glycerol backbone with a fatty acid, usually polyunsaturated, at carbon-2 and a phosphobase (choline, ethanolamine, serine or inositol) at carbon-3 of the glycerol moiety. The fourth class, of which the only representative is sphingo-myelin, contains ceramide linked to phosphocholine through its primary hydroxyl group (Farooqui et al. 1999). The 1,2-diacyl glycerophospholipids include phospha-tidylcholine (PtdCho), phosphatidylethanolamine (PtdEtn), phosphatidylserine (PtdSer), phosphatidylinositol (PtdIns), and phosphatidic acid (PtdOH). The 1-alk-1'-enyl-2-acyl glycerophospholipids include plasmenylethanolamine (PlsEtn) and plasmenylcholine (PlsCho). Small amounts of PakEtn and PakCho, the 1-alkyl-2-acyl types of ethanolamine and choline glycerophospholipids, are also present.

In neural membranes, PtdCho, PtdEtn, PlsEtn and PtdSer contain high levels of 22:6 n-3 acyl groups, whereas PtdIns and PtdOH contain high levels of 20:4 n-6 acyl groups. A high degree of compositional heterogeneity occurs in all subclasses of glycerophospholipids, generally associated with a more diverse acyl group composition at the *sn*-2 position (Nakagawa and Horrocks 1983). Many studies (Horrocks 1989, 1990, Burdge and Postle 1995) suggest that in brain tissue each subclass of glycerophospholipid exists as a heterogeneous mixture of molecular species, and the synthesis of different pools within a glycerophospholipid class appears to be compartmentalized according to the fatty acid composition and the source of the head group (synthesis *de novo* versus modification and interconversion reactions) (Vance 1988). Thus, different pools of molecular species may have different metabolic and physical properties depending upon their location in neural membranes of the developing brain. The purpose of this review article is to describe the occurrence, identification, and role of molecular species of glycerophospholipids in developing and adult brain. It is hoped that this discussion will initiate further studies on the metabolism and role of individual phospholipid molecular species in brain tissue.

2. Glycerophospholipid molecular species in developing brain

2.1. Brain development

Mammalian brain is a complex tissue that undergoes myelination, synaptogenesis, and dendritic arborization during maturation. Being the major constituent of neural membranes, glycerophospholipids are continuously synthesized and catabolized during the brain maturation process. The proportions of ethanolamine plasmalogen are increased from 9% to 19% in rat cerebrum and from 18% to 29% in spinal cord. PtdCho proportions are decreased from 58% to 35% in cerebrum and from 54% to 25% in spinal cord, while the proportions of sphingomyelin increased moderately through maturation of cerebrum and spinal cord (Table 1). The proportions of PtdSer and PtdIns change very little during cerebral development (de Sousa and Horrocks 1979).

In rat brain, total fatty acids increase sharply between 14 and 25 days, reaching 97% of the adult value (38.4±1.2 mg/g wet weight) at 30 days (Patel and Clark 1980). Palmitic, stearic, oleic, arachidonic, and docosahexaenoic acids show similar patterns during development, increasing during the suckling period and reaching 80% or more of the adult value at 25 days of age. Linoleic acid shows a gradual increase per g wet weight of brain during the suckling period, 0 to 21 days, and then a sharp rise to the adult value after weaning (Patel and Clark 1980).

Table 1
Phospholipid composition of rat cerebrum during brain development[a]

Age (days)	PtdEtn	PlsEtn	PtdCho	Sphingomyelin	PtdSer	PtdIns
2	28.6	9.3	57.8	0.7	8.9	4.4
5	20.0	9.7	52.6	0.9	11.9	4.8
7	20.5	9.9	52.0	0.9	12.1	4.6
9	21.2	10.7	51.5	1.0	11.8	3.9
11	21.0	12.3	50.7	1.1	11.2	3.6
13	21.6	12.1	49.2	1.9	11.5	3.2
14	22.1	12.9	46.8	2.6	11.6	3.7
15	23.3	13.1	44.5	3.7	11.9	3.6
20	23.7	14.0	42.8	4.8	11.8	3.3
21	23.5	16.4	40.7	5.1	11.7	3.1
25	23.9	18.3	37.2	5.8	11.3	3.9
30	23.7	18.2	36.1	5.6	11.9	3.9
40	23.6	18.4	34.9	5.1	12.4	3.5
58	24.8	19.0	34.3	5.6	11.6	3.6
90	25.6	18.9	33.5	5.9	11.7	3.5
127	25.4	19.2	33.4	6.7	11.9	3.4

[a] Results are expressed as % of total phosphorus. Modified from de Sousa and Horrocks (1979).

2.2. Species with docosahexaenoate

Studies on the glycerophospholipid molecular species composition in developing fetal guinea-pig brain have shown that an accumulation of docosahexaenoic acid into membrane glycerophospholipid is critical for optimal brain development (Burdge and Postle 1995). This accumulation is maximal during neuritogenesis, which precedes the major growth phase, intense myelination. Total brain PtdCho increases substantially between 40 and 68 days of gestation, corresponding to myelination, while total PtdEtn increases in a biphasic manner. The initial increase is associated with the period of maximal neuritogenesis (25 to 35 days), whereas the second phase corresponds with increase in total brain mass. A significant increase in the level of PlsEtn in late gestation (40 to 68 days) is associated with the intense period of myelination.

During rat brain development, the prenatal docosahexaenoic acid accretion spurt correlates with events of synaptogenesis (Green et al. 1999). The growth cones, which are precursor structures for synapses, also have a high level of molecular species containing docosahexaenoic acid before they develop into mature synapses. This suggests that docosahexaenoic acid plays an important role in synaptogenesis (Martin and Bazan 1992).

2.3. Groups of molecular species

During the maturation of rat brain, 1,2-diacyl-*sn*-glycero-3-phosphocholine (PtdCho), 1,2-diacyl-*sn*-glycero-3-phosphoethanolamine (PtdEtn) and 1-alk-1′-enyl-2-acyl-*sn*-glycero-3-phosphoethanolamine (PlsEtn) comprise approximately 80 to 85% of all glycerophospholipids, while 1-alkyl-2-acyl-*sn*-glycero-3-phosphocholine (PakCho), 1-alk-1′-enyl-2-acyl-*sn*-glycero-3-phosphocholine (PlsCho) and 1-alkyl-2-acyl-*sn*-glycero-3-phosphoethanolamine (PakEtn) account for <3% each (Leray et al. 1990). Leray et al. (1990) divided the molecular species of rat cerebellum into two groups for PtdCho and PtdEtn and three groups for PlsEtn. The first 2 weeks of postnatal life in rat brain tissue are characterized by a large increase in the number and levels of the molecular species of PtdCho, PtdEtn, and alkenyl-GPE belonging to the first group and a much smaller increase in those belonging to the second group. The three molecular species of PlsEtn were absent in the first week of brain development but they increase in content very slowly during the second week after birth. A marked accumulation of glycerophospholipid molecular species of various subclasses occurs during the period of intense myelination and synaptogenesis indicating the formation of biomembranes for neuronal perikarya and glial cells as well as synaptosomal structures. A very slow increase in content of the molecular species of PlsEtn during the first two weeks and their subsequent rapid accumulation during the myelination period suggests that these molecular species are preferentially involved in myelinogenesis and are located in the myelin sheath.

2.4. Molecular species of myelin

The molecular species composition of glycerophospholipid subclasses was determined in myelin membranes from 15, 21, and 90 day-old rat brains. The PtdCho has high proportions of saturated–mono-unsaturated and disaturated species and low proportions of saturated–polyunsaturated species (Leray et al. 1994). During myelin development, the levels of saturated–mono-unsaturated molecular species increased whereas those of the disaturated and saturated–polyunsaturated species decreased in PtdCho. PtdEtn molecular species contain low levels of disaturated and a high proportion of saturated–mono-unsaturated and saturated–polyunsaturated molecular species. The molecular species of PlsEtn include no disaturated species. High proportions of saturated–polyunsaturated, saturated–mono-unsaturated, and dimono-unsaturated species were found. During myelination the molecular species having the n-3 series of PUFA (polyunsaturated fatty acids) are decreased whereas those containing PUFA of the n-6 series are not affected (Leray et al. 1994). A similar distribution of molecular species is also observed in developing guinea-pig and monkey brain (Lin et al. 1990, Burdge and Postle 1995).

The occurrence of so many glycerophospholipid molecular species in myelin membranes complicates the analysis of their function, identification, and roles in developing and adult brain. Their synthesis and deposition in myelin membrane may be controlled not only by their metabolism but also by transport, sorting, and integration mechanisms in oligodendrocytes. This suggestion is supported by studies on the metabolism of molecular species in myelin and alteration of membrane fluidity in adult and developing brain. The occurrence of many enzymic activities such as acyl-CoA: lysophospholipid acyltransferase, long-chain acyl-CoA: synthetase (Vaswani and Ledeen 1987, 1989, phospholipases (Larocca et al. 1987) and choline and ethanolamine phosphotransferases (Ledeen 1984) in myelin suggests that different molecular species in various subclasses turnover with different rates in developing brain.

3. Glycerophospholipid molecular species in adult brain

3.1. Profiles of molecular species

The major classes of glycerophospholipids in adult mammalian brain show very different molecular species profiles (Nakagawa and Horrocks 1983, Lin et al. 1990, Wilson and Bell 1993, Martínez and Mougan 1998). Thus in human white matter, the molecular species of PtdCho and PtdSer contain 85.7% and 82.4%, respectively, of species with only saturated and mono-unsaturated fatty acids. The 18:0–18:1 molecular species dominates in PtdSer and 16:0–18:1 dominates in PtdCho. These molecular species are also abundant in PtdEtn, but in this glycerophospholipid the polyunsaturated species, mainly 18:0–22:6 n-3 and 18:0–20:4 n-6, account for over half of the total species. In contrast, in human white matter PlsEtn, the molecular species 16:0a–18:1 (16:0a is the alk-1′-enyl group with 16 carbon atoms and no double bonds), 18:0a–18:1 and 18:1a–18:1 are present in high amounts and the molecular species containing 20:4 n-6 and 22:6 n-3 are present in low amounts. Two major molecular species of PlsEtn, 16:0a–22:4 n-6 and 18:1a–22:4 n-6, contain 22:4 n-6 as the major polyunsaturated fatty acid (Wilson and Bell 1993). Table 2 shows similar distributions in the diacyl and alkenylacyl types of ethanolamine glycerophospholipids in bovine brain (Lin et al. 1990).

Table 2

The percentage of some polyunsaturated molecular species in ethanolamine glycerophospholipids in bovine brain[a]

Species	PlsEtn	PtdEtn	Species	PlsEtn	PtdEtn
16:0–20:4 n-6	2.8	1.1	18:0–22:6 n-3	8.7	25.5
16:0–22:4 n-6	7.3	1.2	18:1–20:4 n-6	4.7	2.9
16:0–22:6 n-3	4.7	5.5	18:1–22:4 n-6	4.0	1.3
18:0–20:4 n-6	4.8	15.8	18:1–22:6 n-3	2.7	1.7
18:0–22:4 n-6	4.7	8.6			

[a] Modified from Nakagawa and Horrocks (1983).

3.2. Metabolism of molecular species

Among the diacyl type of glycerophospholipids, incorporation of $[^3H]20{:}4$ into rat and mouse brains was highest into the 18:1–20:4 species of PtdCho with a peak value at 60 min after injection (Nakagawa and Horrocks 1986, Horrocks 1989, 1990). The 24-hour values were more than 70% lower, indicating half-lives of 12 hours or less. The 18:0–20:4 species had less than half of the specific radioactivity of the 16:0–20:4 and 18:1–20:4 species at 60 min. The specific radioactivity values of these species of PtdIns were slightly lower than the corresponding values for PtdCho. In addition, the turnover rate of the former was slightly slower. As judged by specific radioactivity values, the metabolism of arachidonic acid in PtdEtn was much more sluggish, particularly in the 18:0–20:4 species, the major component of this glycerophospholipid. The diacyl type of glycerophospholipid was labeled rapidly, but the ether-linked glycerophospholipids had a lag in uptake due to a lack of direct labeling. The latter types were labeled by energy- and CoA-independent transfer of arachidonic acid from PtdCho. The specific radioactivity values of the 16:0a–20:4 and 18:1a–20:4 species of the PlsCho were greater than those for the PtdCho at 30 and 60 min after injection. These values increased between 1 and 24 hours. This may be due to the release of unlabeled arachidonic acid from the PlsCho followed by replenishment from the pool of PtdCho with a very high specific radioactivity.

Part of the injected $[^3H]$arachidonic acid was immediately elongated to $[^3H]$adrenic acid, 22:4 n-6. The uptake of $[^3H]$adrenic acid was much faster into PtdIns than into choline or ethanolamine phospholipids at early times after injection (Horrocks 1989, 1990). There was a considerable turnover of the $[^{14}C]$adrenate in the PtdIns between 1 and 24 hours. Part of this appeared to be transferred to the PakEtn, 1-alkyl-2-acyl-sn-glycero-3-phosphoethanolamines. By 24 hours after injection, the PakEtn showed a preferential uptake of the adrenic acid. All three adrenate molecular species showed a very rapid turnover in the PtdIns. At 30 min the 18:0–22:4 species had less than half the specific radioactivity of the other two molecular species. The turnover of this species was also less than that of the others. The turnover of the 18:1–22:4 species of PtdIns was particularly rapid with an indicated half-life of less than 8 hours. In the PtdCho and the PtdEtn, both the 18:1–22:4 and 16:0–22:4 species showed a good uptake of [^{14}C]adrenic acid and considerable turnover between 1 and 24 hours with half-lives of less than 24 hours. In contrast, the 18:0–22:4 species of PtdCho and PtdEtn gained considerably in specific radioactivity between 60 min and 24 hours. For the PtdCho, this species had the highest specific radioactivity at 24 hours. The specific radioactivity of [^{14}C]22:4 was much higher in the PakEtn than in the PlsEtn. Between 1 and 24 hours, all species of the PlsEtn more than doubled their specific radioactivity whereas only small changes were seen in the PakEtn. At 60 min after injection all of the PakEtn molecular species were at least 7-fold greater in specific radioactivity when compared with the PlsEtn molecular species (Horrocks 1989, 1990). Subcellular distribution studies on the incorporation of [^{3}H]arachidonic acid in rat brain have indicated that this fatty acid selectively incorporates into synaptosomal and microsomal fractions. Lesser incorporation occurs into myelin and mitochondrial fractions. The incorporation of [^{3}H]arachidonic acid into synaptosomal and microsomal compartments is increased by cholinergic stimulation with arecoline (Jones et al. 1996).

Thus in brain tissue the glycerophospholipid molecular species differ in their relative amounts, physical properties, and half lives. For example, the half life of arachidonic acid ($t_{1/2}$) in brain PtdIns molecular species is only about 1 h (Washizaki et al. 1994). It is consistent with the critical role of phosphoinositides in second messenger generation and signal transduction. In contrast, the arachidonic acid $t_{1/2}$ in brain PtdEtn is considerably longer, approximately 24 h, and suggestive of the structural role of these glycerophospholipids in neural membranes and myelin. Studies on the incorporation of docosahexaenoic acid in rat brain subcellular fractions before and after arecoline (a cholinergic agonist) have indicated that this fatty acid preferentially incorporates into membrane glycerophospholipids of synapses. Almost 60% of the label can be localized at synaptosomal membranes in awake control rats (Jones et al. 1997). Cholinergic stimulation by arecoline results in an increase in the incorporation of docosahexaenoic acid into glycerophospholipids of synaptosomal and microsomal fractions. This docosahexaenoic acid incorporation is apparently closely coupled to functional activities of the synaptosomal and microsomal membranes. In contrast, incorporation of plasma [^{3}H]palmitic acid into synaptosomal and microsomal fraction does not change significantly after arecoline treatment (Jones et al. 1997). Like arachidonic acid, docosahexaenoic acid incorporates at the *sn*-2 position of glycerophospholipids whereas palmitic acid is esterified at the *sn*-1 position of glycerol moiety. It must be recognized here that during brain development the acyl group composition of phospholipid molecular species is controlled by the deacylation–reacylation cycle (MacDonald and Sprecher 1991) and is dependent upon the availability of ATP.

The significance of marked differences in the proportion and composition of glycerophospholipid molecular species in brain tissue is not fully understood. However, these differences may be reflected in changes in membrane fluidity and alterations in activities of membrane-bound enzymes, receptor and ion channels (see below). Recent studies have also indicated that glycerophospholipid molecular species may not only fulfill the structural requirements for specific membrane domains for membrane interactions but also serve as resources for the generation of signaling molecules (Litman and Mitchell 1996, Sunshine and McNamee 1994).

4. Methods for separation of molecular species

Many methods have been described for the separation, identification and quantitation of intact glycerophospholipid molecular species from brain tissue (Nakagawa and Horrocks 1983, Horrocks 1989, 1990, Wiley et al. 1992, Ma and Kim 1995, McHowat et al. 1997) and have been reviewed in detail elsewhere (Olsson and Salem Jr 1997). In this review we will not discuss the details of methodology used for the separation of molecular species but rather focus on the detection and quantitation of molecular species. In general, it is a good idea to avoid derivatization because it increases the risk for the introduction of artifacts. The chromatographic separation of glycerophospholipid molecular species requires high resolving power, as there are only subtle distinctions between many molecular species, such as the presence of one extra double bond or two extra carbon atoms. More than 100 molecular species have been separated from rat liver and heart phospholipid classes (Wiley et al. 1992). Vecchini et al. (1997) described a novel method for the separation of molecular species. In this method diacylglycerols obtained by hydrolyzing phospholipids by phospholipase C can be separated into the diacyl-, alkylacyl-, and alkenylacyl- subclasses by HPLC. The molecular species of underivatized diacylglycerol can be separated by reverse-phase octadecyl-silica column chromatography. Quantitation of molecular species is usually performed with a UV detector (Wiley et al. 1992, McHowat et al. 1996). The response per molecule is dependent upon the degree of unsaturation of the aliphatic groups. For this detection, making careful calibration for each component is necessary. Light scattering is another excellent choice for the detection of molecular species. In contrast to UV detection, light scattering can be more sensitive and the response is independent of the number of double bonds in the molecular species (Guiochon et al. 1988, Brouwers et al. 1998, 1999).

The interaction between head group and stationary phase can be eliminated by the addition of quaternary amine salts, which compete with the head group for the interaction sites on the reverse-phase column. Mass spectrometry is another method of detection and quantitation that has been used in several studies (Ma and Kim 1995). Fluorescent derivatives of glycerophospholipid molecular species have also been analyzed by interfacing HPLC separation with a fluorescence detector (Rastegar et al. 1990). In this procedure the formation of fluorescent micelles enables the detection of fully saturated PtdCho species, but the mixing of the column effluent with the fluorescent 1,6-diphenyl-1,3,5-hexatriene results in loss of resolution (Postle 1987). Thus, careful consideration must be given to the solvents and detection procedures.

5. Roles of molecular species in brain tissue

Assignment of a specific role to a particular glycerophospholipid molecular species in developing brain is not possible now. The identification of specific changes in the content of specific molecular species at a defined point in brain development and information about their half-life in various subcellular structures may provide a basis for defining the role of molecular species in maintenance of membrane function in brain tissue. In developing and adult brain, glycerophospholipid molecular species containing

Table 3
Possible roles of arachidonate and docosahexaenoate containing molecular species of neural membrane glycerophospholipids

Fatty acid at sn-2 position	Role	Reference
Arachidonate	Signal transduction (generation of second messengers)	(Farooqui et al. 1997)
	Gene expression and differentiation	(Farooqui et al. 1996)
	Regulation of enzymic activities (Na^+, K^+-ATPase and PKC)	(Bourre et al. 1989, Slater et al. 1996)
Docosahexaenoate	Visual function (activation of rhodopsin)	(Mitchell and Litman 1999)
	Learning and memory	
	Membrane fusion	(Stillwell et al. 1993)
	Prevention of apoptosis	(Rotstein et al. 1997)
	Activation of DAG-kinase	(Vaidyanathan et al. 1994)

polyunsaturated fatty acids may be involved in the regulation of neural membrane fluidity and permeability (Litman and Mitchell 1996, Draz and Holte 1993) and in the activation of rhodopsin during visual function and learning and memory processes (Sinclair et al. 1997, Salem Jr 1989, Mitchell and Litman 1998). The addition of PtdCho containing docosahexaenoic acid to both synthetic and biomembranes enhances membrane fusion and permeability (Ehringer et al. 1990, Stillwell et al. 1993, 1997). Molecular species of glycerophospholipids containing docosahexaenoic acid interact poorly with cholesterol (Jenski et al. 1999) and induce distinct membrane microdomains. These affect the domain distribution of membrane-bound proteins, which are reflected in alterations in their biological and enzymic activities (Table 3).

Thus, the replacement of docosahexaenoic acid with α-linolenic acid results in significant reduction of Na^+, K^+-ATPase activity in the nerve terminals of rat brain (Bourre et al. 1989, Gerbi et al. 1998). In nerve terminals, the regulation of this enzyme by molecular species plays an important role in Na^+ and K^+ transport by maintaining the ionic gradient necessary for both cellular metabolism and nerve excitability. A decrease in transport rate may result in an increase in the intracellular concentration of Na^+ and a corresponding decrease in membrane Na^+ potential. In the nerve, this process can induce a reduction in the conduction velocity. Glycerophospholipid molecular species may also be involved in activating and stabilizing enzymes involved in intracellular signaling. For example, PtdSer molecular species are required for protein kinase C (PKC) activation. PKC is less active in the presence of those molecular species of PtdSer that are rich in n-3 polyunsaturated fatty acids compared to those species with n-3 polyunsaturated-poor PtdSer molecular species, irrespective of the fatty acid composition of diacylglycerols in the membranes. This suggests that specific molecular species of PtdSer are needed for the optimal activation of PKC during brain development (Slater et al. 1996). Thus in neural membranes the molecular species composition of the glycerophospholipid substrate pool may modulate neural cell responses to agonists, growth factors and neurotransmitters. Another possible role of molecular species containing arachidonic acid or docosahexaenoic acid may be to provide second messengers (eicosanoids and docosanoids) to the developing and adult brain for signal transduction processes (Jones et al. 1997, Litman and Mitchell 1996, Yokota et al. 1999). Preliminary evidence indicates that at least some docosanoids are antagonistic to eicosanoids (VanRollins 1991). The number of unsaturated carbon–carbon bonds in the aliphatic groups at the sn-1 and sn-2 position of glycerophospholipid molecular species is an important factor in determining phase transition temperature, lateral diffusion velocity, and thickness of neural membranes. The molecular species impart these properties to membranes and may play important roles in endo and exocytosis, lipid sorting, and membrane fusion. Molecular species containing docosahexaenoic acid may also be involved in the prevention of apoptotic cell death in retinal photoreceptor cells (Rotstein et al. 1997).

Molecular species containing arachidonic acid are rapidly hydrolyzed by cytosolic phospholipase A_2 (cPLA$_2$), an enzyme that specifically liberates arachidonic acid and may be involved in apoptotic cell death during brain development and neurodegeneration (Farooqui et al. 1997). Glycerophospholipid molecular species that have docosahexaenoic acid may prevent apoptotic cell death in photoreceptor cells (Rotstein et al. 1997) because cPLA$_2$ does not act on molecular species that have docosahexaenoic acid at the sn-2 position (Shikano et al. 1994). Thus, the alteration of brain glycerophospholipid composition by diets rich in docosahexaenoic acid may protect against apoptotic cell death that occurs during acute neural trauma and neurodegenerative disease.

6. Conclusion

Molecular species with a high degree of heterogeneity for various subclasses are present in adult and developing brain. Part of the heterogeneity is due to the diverse acyl chain composition at the sn-2 position. The molecular species of brain tissue phospholipid subclasses can be separated by silicic acid HPLC and reverse-phase HPLC on octadecyl-silica columns. A marked accumulation of specific glycerophospholipid molecular species occurs during the period of intense myelination and synaptogenesis. In myelin, the proportions of docosahexaenoic acid and arachidonic acid are lower, but those of adrenic and oleic acids are high. The mature brain contains high proportions of polyunsaturated fatty acids in the alkenylacyl glycerophospholipids, the plasmalogens. Glycerophospholipid molecular species may be involved in regulation of neural membrane fluidity, permeability, visual function, and learning skills. Molecular species regulate the activities of many enzymes, including Na$^+$, K$^+$-ATPase and protein kinase C. The molecular species also provide second messengers to the brain tissue for receptor function. Molecular species containing docosahexaenoic acid may also be involved in prevention of apoptotic cell death in retinal photoreceptor cells.

References

Bourre, J.-M., Francois, M., Youyou, A., Dumont, O., Piciotti, M., Pascal, G. and Durand, G. (1989) The effects of dietary α-linolenic acid on the composition of nerve membranes, enzymatic activity, amplitude of electrophysiological parameters, resistance to poisons and performance of learning tasks in rats. J. Nutr. 119, 1880–1892.

Brouwers, J.F.H.M., Gadella, B.M., Van Golde, L.M.G. and Tielens, A.G.M. (1998) Quantitative analysis of phosphatidylcholine molecular species using HPLC and light scattering detection. J. Lipid Res. 39, 344–353.

Brouwers, J.F.H.M., Vernooij, E.A.A.M., Tielens, A.G.M. and Van Golde, L.M.G. (1999) Rapid separation and identification of phosphatidylethanolamine molecular species. J. Lipid Res. 40, 164–169.

Burdge, G.C. and Postle, A.D. (1995) Phospholipid molecular species composition of developing fetal guinea pig brain. Lipids 30, 719–724.

de Sousa, B.N. and Horrocks, L.A. (1979) Development of rat spinal cord II. Comparison of lipid composition with cerebrum. Dev. Neurosci. 2, 122–128.

Draz, E.E. and Holte, S. (1993) The molecular spring model for the function of docosahexaenoic acid (22:6 n-3) in biological membranes. In: A.J. Sinclair and R.A. Gibson (Eds.), Essential Fatty Acids and Eicosanoids: Invited Papers from the Third International Congress, American Oil Chemists' Society Press, Champaign, IL, pp. 122–127.

Ehringer, W., Belcher, D., Wassall, S.R. and Stillwell, W. (1990) A comparison of the effects of linolenic (18:3 omega 3) and docosahexaenoic (22:6 omega 3) acids on phospholipid bilayers. Chem. Phys. Lipids 54, 79–88.

Farooqui, A.A., Rosenberger, T.A. and Horrocks, L.A. (1996) Arachidonic acid, neurotrauma and neurodegenerative diseases. In: D.I. Mostofsky and S. Yehuda (Eds.), Fatty Acids: Biochemistry and Behavior, Humana Press, Totowa, NJ.

Farooqui, A.A., Yang, H.C., Rosenberger, T.A. and Horrocks, L.A. (1997) Phospholipase A_2 and its role in brain tissue. J. Neurochem. 69, 889–901.

Farooqui, A.A., Yeo, Y.K. and Horrocks, L.A. (1999) Glycerophospholipids. In: MacMillan Reference Limited (Ed.), Encyclopaedia of Life Sciences, Stockton Press, London, in press.

Gerbi, A., Maixent, J.M., Barbey, O., Jamme, I., Pierlovisi, M., Coste, T., Pieroni, G., Nouvelot, A., Vague, P. and Raccah, D. (1998) Alterations of Na, K-ATPase isoenzymes in the rat diabetic neuropathy: protective effect of dietary supplementation with n-3 fatty acids. J. Neurochem. 71, 732–740.

Green, P., Glozman, S., Kamensky, B. and Yavin, E. (1999) Developmental changes in rat brain membrane lipids and fatty acids: the preferential prenatal accumulation of docosahexaenoic acid. J. Lipid Res. 40, 960–966.

Guiochon, G., Moysan, A. and Holley, C. (1988) Influence of various parameters on the response factors of the evaporative light scattering detector for a number of non-volatile compounds. J. Liquid Chromatogr. 11, 2547–2570.

Horrocks, L.A. (1989) Sources of brain arachidonic acid uptake and turnover in glycerophospholipids. Ann. N.Y. Acad. Sci. 559, 17–24.

Horrocks, L.A. (1990) Metabolism of adrenic and arachidonic acids in nervous system phospholipids. In: I. Hanin and G. Pepeu (Eds.), Phospholipids: Biochemical Pharmaceutical and Analytical Considerations, Plenum Press, New York, pp. 51–58.

Jenski, L.J., Caldwell, L.D., Jiricko, P., Scherer, J.M. and Stillwell, W. (1999) DHA-facilitated membrane protein clustering and T cell activation. In: R.A. Riemersma, R. Armstrong, R.W. Kelly and R. Wilson (Eds.), Essential Fatty Acids and Eicosanoids: Invited Papers from the Fourth International Congress, American Oil Chemists' Society Press, Champaign, IL, pp. 191–196.

Jones, C.R., Arai, T., Bell, J.M. and Rapoport, S.I. (1996) Preferential in vivo incorporation of [^{3}H]arachidonic acid from blood into rat brain synaptosomal fractions before and after cholinergic stimulation. J. Neurochem. 67, 822–829.

Jones, C.R., Arai, T. and Rapoport, S.I. (1997) Evidence for the involvement of docosahexaenoic acid in cholinergic stimulated signal transduction at the synapse. Neurochem. Res. 22, 663–670.

Larocca, J.N., Cervone, A. and Ledeen, R.W. (1987) Stimulation of phosphoinositide hydrolysis in myelin by muscarinic agonist and potassium. Brain Res. 436, 357–362.

Ledeen, R.W. (1984) Lipid-metabolizing enzymes of myelin and their relation to the axon. J. Lipid Res. 25, 1548–1554.

Leray, C., Pelletier, A., Massarelli, R., Dreyfus, H. and Freysz, L. (1990) Molecular species of choline and ethanolamine phospholipids in rat cerebellum during development. J. Neurochem. 54, 1677–1681.

Leray, C., Sarlième, L.L., Dreyfus, H., Massarelli, R., Binaglia, L. and Freysz, L. (1994) Molecular species of choline and ethanolamine glycerophospholipids in rat brain myelin during development. Lipids 29, 77–81.

Lin, D.S., Connor, W.E., Anderson, G.J. and Neuringer, M. (1990) Effects of dietary n-3 fatty acids on the phospholipid molecular species of monkey brain. J. Neurochem. 55, 1200–1207.

Litman, B.J. and Mitchell, D.C. (1996) A role for phospholipid polyunsaturation in modulating membrane protein function. Lipids 31, S193–S197.

Ma, Y.-C. and Kim, H.-Y. (1995) Development of the on-line high-performance liquid chromatography/thermospray mass spectrometry method for the analysis of phospholipid molecular species in rat brain. Anal. Biochem. 226, 293–301.

MacDonald, J.I.S. and Sprecher, H. (1991) Phospholipid fatty acid remodeling in mammalian cells. Biochim. Biophys. Acta 1084, 105–121.

Martin, R.E. and Bazan, N.G. (1992) Changing fatty acid content of growth cone lipids prior to synaptogenesis. J. Neurochem. 59, 318–325.

Martínez, M. and Mougan, I. (1998) Fatty acid composition of human brain phospholipids during normal development. J. Neurochem. 71, 2528–2533.

McHowat, J., Jones, J.H. and Creer, M.H. (1996) Quantitation of individual phospholipid molecular species by UV absorption measurements. J. Lipid Res. 37, 2450–2460.

McHowat, J., Jones, J.H. and Creer, M.H. (1997) Gradient elution reversed-phase chromatographic isolation of individual glycerophospholipid molecular species. J. Chromatogr. B 702, 21–32.

Mitchell, D.C. and Litman, B.J. (1998) Molecular order and dynamics in bilayers consisting of highly polyunsaturated phospholipids. Biophys. J. 74, 879–891.

Mitchell, D.C. and Litman, B.J. (1999) Docosahexaenoic acid-containing phospholipids optimally promote rhodopsin activation. In: R.A. Riemersma, R. Armstrong, R.W. Kelly and R. Wilson (Eds.), Essential Fatty Acids and Eicosanoids: Invited Papers from the Fourth International Congress, American Oil Chemists' Society Press, Champaign, IL, pp. 154–158.

Nakagawa, Y. and Horrocks, L.A. (1983) Separation of alkenylacyl, alkylacyl, and diacyl analogues and their molecular species by high performance liquid chromatography. J. Lipid Res. 24, 1268–1275.

Nakagawa, Y. and Horrocks, L.A. (1986) Different metabolic rates for arachidonoyl molecular species of ethanolamine phosphoglycerides in rat brain. J. Lipid Res. 27, 629–636.

Olsson, N.U. and Salem Jr, N. (1997) Molecular species analysis of phospholipids. J. Chromatogr. B. 692, 245–256.

Patel, T.B. and Clark, J.B. (1980) Comparison of the development of the fatty acid content and composition of the brain of a precocial species (guinea pig) and a non-precocial species (rat). J. Neurochem. 35, 149–154.

Postle, A.D. (1987) Method for the sensitive analysis of individual molecular species of phosphatidylcholine by high-performance liquid chromatography using post-column fluorescence detection. J. Chromatogr. 415, 241–251.

Rastegar, A., Pelletier, A., Duportail, G., Freysz, L. and Leray, C. (1990) Sensitive analysis of phospholipid molecular species by high-performance liquid chromatography using fluorescent naproxen derivatives of diacylglycerols. J. Chromatogr. 518, 157–165.

Rotstein, N.P., Aveldaño, M.I., Barrantes, F.J., Roccamo, A.M. and Politi, L.E. (1997) Apoptosis of retinal photoreceptors during development in vitro: Protective effect of docosahexaenoic acid. J. Neurochem. 69, 504–513.

Salem Jr, N. (1989) Omega-3 fatty acids: molecular and biochemical aspects. In: G.A. Spiller and J. Scala (Eds.), New Protective Roles for Selected Nutrients, Alan R. Liss, New York, pp. 109–228.

Shikano, M., Masuzawa, Y., Yazawa, K., Takayama, K., Kudo, I. and Inoue, K. (1994) Complete discrimination of docosahexaenoate from arachidonate by 85 kDa cytosolic phospholipase A_2 during the hydrolysis of diacyl- and alkenylacylglycerophosphoethanolamine. Biochim. Biophys. Acta Lipids Lipid Metab. 1212, 211–216.

Sinclair, A.J., Weisinger, H.S. and Vingrys, A.J. (1997) Neural function following dietary n-3 fatty acid depletion. In: S. Yehuda and D.I. Mostofsky (Eds.), Handbook of Essential Fatty Acid Biology, Humana Press, Totowa, NJ, pp. 201–214.

Slater, S.J., Kelly, M.B., Yeager, M.D., Larkin, J., Ho, C. and Stubbs, C.D. (1996) Polyunsaturation in cell membranes and lipid bilayers and its effects on membrane proteins. Lipids 31, S189–S192.

Stillwell, W., Ehringer, W. and Jenski, L.J. (1993) Docosahexaenoic acid increases permeability of lipid vesicles and tumor cells. Lipids 28, 103–108.

Stillwell, W., Jenski, L.J., Crump, F.T. and Ehringer, W. (1997) Effect of docosahexaenoic acid on mouse mitochondrial membrane properties. Lipids 32, 497–506.

Sunshine, C. and McNamee, M.G. (1994) Lipid modulation of nicotinic acetylcholine receptor function: the role of membrane lipid composition and fluidity. Biochim. Biophys. Acta 1191, 59–64.

Vaidyanathan, V.V., Rao, K.V.R. and Sastry, P.S. (1994) Regulation of diacylglycerol kinase in rat brain membranes by docosahexaenoic acid. Neurosci. Lett. 179, 171–174.

Vance, J.E. (1988) Compartmentalization of phospholipids for lipoprotein assembly on the basis of molecular species and biosynthetic origin. Biochim. Biophys. Acta 963, 70–81.

VanRollins, M. (1991) Cytochrome P-450 metabolites of docosahexaenoic acid inhibit platelet aggregation without affecting thromboxane production. In: A.P. Simopoulos, R.R. Kifer and C. Barlow (Eds.), Health Effects of ω3 Polyunsaturated Fatty Acids in Seafoods, Karger, Basel, pp. 502–502.

Vaswani, K.K. and Ledeen, R.W. (1987) Long-chain acyl-coenzyme A synthetase in rat brain myelin. J. Neurosci. Res. 17, 65–70.

Vaswani, K.K. and Ledeen, R.W. (1989) Purified rat brain myelin contains measurable acyl-CoA:lyso-phospholipid acyltransferase(s) but little, if any, glycerol-3-phosphate acyltransferase. J. Neurochem. 52, 69–74.

Vecchini, A., Panagia, V. and Binaglia, L. (1997) Analysis of phospholipid molecular species. Mol. Cell Biochem. 172, 129–136.

Washizaki, K., Smith, Q.R., Rapoport, S.I. and Purdon, A.D. (1994) Brain arachidonic acid incorporation and precursor pool specific activity during intravenous infusion of unesterified [^{3}H]arachidonate in the anesthetized rat. J. Neurochem. 63, 727–736.

Wiley, M.G., Przetakiewicz, M., Takahashi, M. and Lowenstein, J.M. (1992) An extended method for separating and quantitating molecular species of phospholipids. Lipids 27, 295–301.

Wilson, R. and Bell, M.V. (1993) Molecular species composition of glycerophospholipids from white matter of human brain. Lipids 28, 13–17.

Yokota, K., Tsuruhami, K., Nishimura, K., Nagaya, T., Jisaka, M. and Takinami, K. (1999) Modification of membrane phospholipids with n-6 and n-3 essential fatty acids regulates the gene expression of prostaglandin endoperoxide synthase isoforms upon agonist-stimulation. In: R.A. Riemersma, R. Armstrong, R.W. Kelly and R. Wilson (Eds.), Essential Fatty Acids and Eicosanoids: Invited Papers from the Fourth International Congress, American Oil Chemists' Society Press, Champaign, IL, pp. 68–73.

E.R. Skinner (Ed.), *Brain Lipids and Disorders in Biological Psychiatry*
© 2002 Elsevier Science B.V.   All rights reserved

CHAPTER 9

Polyunsaturated fatty acids, brain phospholipids and the fetal alcohol syndrome

Graham C. Burdge

Institute of Human Nutrition, University of Southampton, Southampton, UK

Anthony D. Postle

Department of Child Health, University of Southampton, Southampton, UK

1. Introduction

Chronic consumption of large amounts of alcohol by pregnant women carries an increased risk of their infants being born with severe neurological damage and impaired physical development. The term fetal alcohol syndrome (FAS) was coined in the early 1970s to describe this condition (Jones et al. 1973), although many previous reports are to be found in the literature. The estimated world-wide prevalence of FAS is reported as 1.9/1000 live births, but may be greater in specific ethnic or social groups. For example, the prevalence amongst some tribes of native Americans may be as high as 19.5/1000 live births (Abel and Sokol 1987). FAS is characterized by neurological and facial dysmorphia which may be accompanied, depending on severity, by structural abnormalities and dysfunction of the major organs. Although the severity of dysmorphia may decrease with increasing age of the child, ethanol-induced developmental and functional deficits to the nervous system are permanent. Such neurological damage may result in marked microcephaly, hypotonia and hyperactivity accompanied by substantially reduced IQ and learning abilities (Henderson et al. 1981).

The incidence of FAS in children from alcoholic mothers is highly variable, and chronic severely alcoholic women are sub-fertile and experience a higher frequency of spontaneous abortion. About 30% of their viable offspring have been estimated to develop the recognised clinical symptoms of FAS, and the severity of both neurological and somatic developmental abnormalities appears to be directly related to the amount of ethanol consumed. For example, while chronic maternal consumption of large amounts of ethanol throughout pregnancy may result in the most severe form of FAS, consumption of only 2U ethanol/day has been associated with reduced IQ at 4 years of age (Streissguth et al. 1989). Such sensitivity to teratogenic action may not be surprising in the context of the unusual complexity and prolonged development of the human brain. The processes of neuronal differentiation and neuritogenesis in human fetal brain development begin in the second trimester and continue to at least 18 months of post-natal age. By contrast, neuronal differentiation of rat brain is predominantly post-natal, while this process is essentially complete at birth for the guinea pig. This considerable variation between animal species in the timing of neuritogenesis is one critical factor in the interpretation of animal models of brain development in general and of FAS in particular.

The precise effects of ethanol exposure on the development of brain structure are not clear. The decrease in brain mass and associated microcephaly suggests decreased numbers of cells and interneurone connections. This is supported by histological examination of fetal mouse brain exposed to ethanol *in utero* which showed decreased formation of interneurone connections in specific brain regions (Ledig et al. 1991). Although it is probable that inadequate nutrition and the possible use of cigarettes or other drugs may contribute to the overall severity of the syndrome, there is strong evidence that ethanol is the principal causative agent in FAS. However, the precise biochemical mechanism by which ethanol exposure results in impaired development is not known.

2. Polyunsaturated fatty acids and brain development

As lipid comprises some 70% of the dry weight of the brain, it is perhaps not surprising that many aspects of lipid metabolism are critical for brain development as well as for brain function. In this context, the incorporation of long-chain polyunsaturated fatty acids (LCPUFA) into neuronal cell membranes, particularly into the growth cone and developing synapse, is an integral component of neuronal differentiation. The phosphatidylethanolamine (PE) fraction of neuronal lipid is uniquely enriched in PUFA of the n-3 series, in particular docosahexaenoic acid (22:6 n-3, DHA), and considerable evidence suggests that DHA-containing PE is important for neuronal function. Much of this evidence comes from studies where n-3 PUFA supply to the fetus and neonate has been inadequate, either due to nutritional restriction in animal models or to preterm delivery of human infants. Many of these studies have concentrated on the role of n-3 PUFA in retinal development and function, both because DHA is highly enriched in rod outer membrane PE and because assessment of retinal function is easier than of cognitive function. For instance, offspring of monkeys fed a diet throughout pregnancy that was deficient in n-3 PUFA showed decreased brain and retinal DHA contents accompanied by a reduced retinal response and possible altered neurological function, indicated by significant polydipsia (Neuringer et al. 1989). These deficits in neurological function appeared to be irreversible after the initial neonatal period.

Much research emphasis has been directed towards the consequences of preterm birth, both on the neurological function and on the phospholipid composition of neonatal brain. Placental nutrition to the fetus is characterised by preferential delivery of LCPUFA (DHA, arachidonic acid, 20:4 n-6) to fetal tissues in preference to supply of their precursor fatty acids (α-linolenic acid, 18:3 n-3, and linoleic acid, 18:2 n-6). This directed lipid supply from the maternal circulation to the fetus is quantitatively greatest over the last trimester of pregnancy. Preterm delivery abruptly terminates placental nutrition, and the neonate then has to rely on its own ability to elongate and desaturate the fatty acid substrates for LCPUFA synthesis. As these activities are low in neonatal human liver, preterm delivery effectively initiates a phase of relative essential fatty acid deficiency. As a result, one consequence of preterm birth is a deficit in accumulation of n-3 PUFA into neurological tissues. For instance, human infants born preterm and fed milk formula lacking n-3 PUFA exhibited reduced visual acuity and evoked potential responses compared with similar infants fed breast milk containing DHA (Uauy et al. 1990). This result was essentially the same as that observed in n-3 PUFA-deprived monkeys (Neuringer et al. 1989). Concerns about these neurological consequences of preterm delivery have resulted in the introduction throughout Europe of n-3 PUFA supplements for all formula feeds designed for preterm infants. A number of studies have now demonstrated that such supplementation of preterm infant formula can reduce the severity of the deficit in visual function (Uauy et al. 1990), although evidence to support a more general beneficial effect on brain function is less clear. The concept that neurological function is related to membrane n-3 PUFA content is supported further by the observation that electrical response in guinea-pig retina increased with increasing DHA content, although this relationship decreased at high DHA levels (Weisinger et al. 1996).

The relationship between DHA availability during development and brain function is difficult to assess in humans. A number of studies have compared the effects on parameters of intellectual function of feeding preterm infants with either milk formula lacking n-3 PUFA or with breast milk. Such studies have produced conflicting evidence about the precise role of n-3 PUFA in human brain function. However, a possible causal relationship has been suggested in children born preterm between reduced IQ at eight years of age and feeding milk formula, presumably lacking n-3 PUFA, during the neonatal period (Lucas et al. 1992). In addition, reduced persistence of moderate neurological dysfunction at nine years has been observed in children who were born preterm and fed formula feed compared with those who received breast milk (Lanting et al. 1994).

3. Alcohol and brain phospholipid composition

Adequate accumulation of n-3 PUFA into neural membranes during fetal development appears, therefore, to have long-term consequences for neurological function. One hypothesis that has been proposed to explain, at least in part, the harmful effects of chronic prenatal ethanol exposure on subsequent neurological function is that ethanol impairs accumulation of DHA into developing brain phospholipids. Such a view is consistent with reports of the effects of ethanol administered to adult animals either by inhalation (La Droitte et al. 1984, Littleton and John 1977, Littleton et al. 1979) or liquid diet (Corbett et al. 1992, Gustavsson and Alling 1989), which showed a decreased content of DHA in brain phospholipid. It is possible that this ethanol-induced decrease in the DHA content of brain PE may be due to enhanced turnover secondary to increased lipid peroxidation.

The pattern of accumulation of DHA into aminophospholipids of the fetal and neonatal brain is temporally co-ordinated with the timing of the establishment of interneurone connections. As mentioned above, both processes start in the human fetal brain at about 16 weeks after conception and continue postnatally until at least 6 months after birth (Clandinin et al. 1980a,b). These data, together with the observation that synapses are enriched in DHA (Salem 1989), suggest that accumulation of DHA into neural membranes is associated with the principal period of neuritogenesis. Consequently, if maternal alcohol consumption during pregnancy impairs both the accumulation of DHA in synapse phospholipids and the process of neuritogenesis, this could be one mechanism to explain the irreversible neurological damage characteristic of FAS.

4. Animal models of fetal alcohol syndrome

As it is not possible to investigate such proposed mechanisms of FAS in human development, the use of suitable animal models has proved essential. Given the complexity and prolonged duration of human brain development compared with all other animal species, the choice of animal model to study has inevitably been a compromise dictated by cost and by patterns both of placental lipid supply to the fetus and of brain development. Many studies have been performed in rats, mice and guinea pigs, but as yet not in primates.

Since FAS is due to prenatal exposure of developing neural tissue to ethanol, the major characteristic for an animal model is that the majority of neurite formation and DHA accumulation both occur before birth. In this context, animal species such as the rat and mouse are poor models for FAS in human infants. Brain maturation and neuritogenesis are essentially postnatal in these species, and the major portion of DHA supply to the brain also occurs in the first week after birth (Sinclair and Crawford 1972). There are also considerable differences in lipid transport between the rat and human placenta, with the rat placenta being relatively impermeable to LCPUFA. Consequently, while ethanol administration to pregnant rats can lead to significant teratogenic effects on rat pups, these are essentially gross somatic effects rather than primary neurological actions characteristic of FAS. In contrast, both brain development (Dobbing and Sands 1979) and accumulation of DHA into brain phospholipids are almost completely prenatal events in the guinea pig, and properties of lipid transport are similar for the guinea-pig and human placentas. Obviously, guinea-pig brain development is comparatively simple, and this animal would not be a good model for the complexity of brain development in the human infant. However, it is potentially a very good model to study the effects of prenatal ethanol exposure on the processes of delivery of LCPUFA to the fetal brain and of neuritogenesis.

5. Fetal alcohol syndrome in the developing guinea pig

Consequently, we have developed a guinea-pig model to study fetal lipid nutrition and brain development. Using this model, we have characterized mechanisms responsible for the directed supply of DHA and other PUFAs to brain PE, and have then investigated the effects of chronic maternal ethanol exposure on brain phosphatidylcholine (PC) and PE molecular species concentration. Finally, we have evaluated the potential effect of dietary supplementation with tuna fish oil, which was enriched in DHA, on mitigating the pathology of FAS in the fetal guinea pig.

5.1. Brain phospholipids of the developing guinea pig

HPLC analysis of fetal guinea-pig brain PC and PE (Burdge and Postle 1995a) showed distinct molecular species contents. Brain PC was characterized predominantly by saturated and mono-unsaturated species, principally PC16:0/16:0 and PC16:0/18:1, while PE contained mainly polyunsaturated species. In addition, term (68 days gestation) fetal guinea-pig brain showed a similar composition to both human (Wilson and Bell 1993) and rat (Hullin et al. 1989) brain PC and PE. These observations support the concept that optimal neurological function requires a precise molecular composition of membrane phospholipid. Measurement of fetal guinea-pig brain PC and PE composition at 25, 35, 40 and 68 (term) days gestation ($n = 6$/gestational age) showed progressive changes to the concentration of selected individual molecular species. The concentration of the major fetal guinea-pig brain PC species PC16:0/16:0 and PC16:0/18:1 doubled between day 25 and term, while PC16:0/18:0 and PC18:0/18:1 contents increased between days 35 and 40, and 40 and term, respectively (Burdge and Postle 1995a). Maturation of fetal guinea-pig PE composition was characterized by initial incorporation of DHA into sn-1 16:0 species between 25 and 40 days gestation, and into sn-1 18:0 and 18:1 n-9 species between 40 days and term (Fig. 1). These results suggest that the composition of developing brain PC and PE is regulated closely and that there is a requirement for a specific membrane phospholipid composition at precise time points in gestation. Conversely, failure to achieve an appropriate membrane composition at a particular developmental time point may result in a permanent functional or structural deficit. Neuritogenesis in fetal guinea-pig brain occurs mainly between 25 and 40 days gestation, with onset of electrical activity at about 45 days. Since in both adult rat (Samborski et al. 1990) and

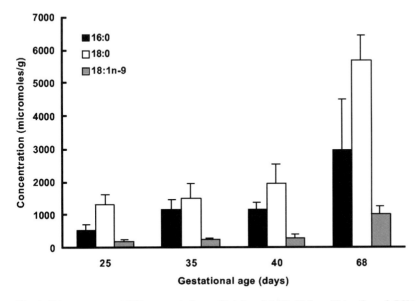

Fig. 1. Values are mean±S.D. concentrations of total sn-1 16:0, 18:0 or 18:1 n-9, sn-2 DHA fetal guinea-pig brain phosphatidylethanolamine molecular species.

fetal guinea pig (Burdge et al. 1993) liver PE16:0/22:6 is turned over more rapidly than PC18:0/22:6, initial assimilation of DHA into fetal guinea-pig brain PE *sn*-1 16:0 species may reflect a relatively short-lived pool to support neurite formation and out-growth. In contrast, accumulation of DHA into *sn*-1 18:0 and 18:1 n-9 PE species may represent a more stable pool consistent with establishment of long-term inter-neural connections.

5.2. Maternal ethanol feeding and fetal guinea-pig development

Adult guinea pigs were fed a high dose of ethanol (6 g/kg/day) before and throughout pregnancy. This feeding regimen produced marked changes to brain phospholipid composition at term compared with chow-fed controls ($n = 6$ fetuses/group) (Burdge and Postle 1995b). Fetal brain after ethanol exposure had significantly greater concentrations of PC16:0/16:0, PC16:0/18:1 and PC18:0/18:1, and decreased contents of DHA-containing species. Furthermore, all PE species containing DHA and arachidonic acid (20:4 n-6) were decreased in ethanol-exposed term fetal brain (Fig. 2). In particular, PE18:0-alkyl/22:6-acyl was absent, while PE16:0/18:1 was present only in fetuses exposed to ethanol. These data indicate that prenatal exposure to ethanol impaired the maturation-associated programmed changes to brain phospholipid composition observed in the control animals.

One possible explanation for decreased accumulation of DHA into fetal brain is impaired supply of DHA or its precursor from the mother. However, analysis of maternal liver and plasma PC compositions showed that the pregnancy-associated increase in PC16:0/22:6 concentration which may represent a means of increasing DHA availability to the fetus (Burdge and Postle 1994, Burdge et al. 1994, Postle et al. 1995) was not altered significantly by ethanol consumption (Burdge et al. 1996). Together with the differential effect of ethanol exposure on the concentrations of individual DHA-containing molecular species, these data suggest that the effect of ethanol is primarily a direct action on the fetal brain. Such changes to membrane composition must be the net consequence of modifications to the specificity or rate of either the synthesis or turnover of individual molecular species of phospholipids, but it is not possible to distinguish these processes by mass measurements alone. Maternal ethanol exposure caused severe effects on aspects of the physiological development of the fetal guinea pig. Gross motor changes to ethanol-exposed guinea-pig pups delivered at term included impaired hind limb function and loss of righting reflex. Such changes may be partly explained by altered neural function.

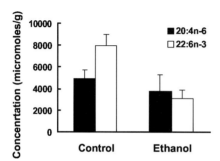

Fig. 2. Bars represent mean±S.D. phosphatidylethanol-amine arachidonic acid (20:4 n-6) and DHA concentrations from control or ethanol-exposed fetal guinea-pig brain at term.

5.3. Dietary supplementation and fetal alcohol syndrome in the fetal guinea pig

The apparent association between ethanol exposure and decreased brain phospholipid DHA content presented the possibility that maternal dietary supplementation with DHA could reduce the severity of the effects of ethanol by increasing DHA availability to the fetus. To test this hypothesis, adult female guinea pigs were fed for fourteen days before and throughout pregnancy one of four diets: chow, chow with ethanol (6 g/kg/day), chow with DHA-enriched (26% total fatty acids) tuna oil to provide 130 mg DHA per day, or chow, tuna oil and ethanol. Guinea-pig pups were delivered at term and brain PC and PE fatty acid composition determined (Burdge et al. 1997). The offspring of mothers fed tuna oil and chow alone did not show any significant difference in brain phospholipid DHA content compared with controls. This suggests that the amount of DHA in fetal guinea-pig brain phospholipids is tightly regulated, and that the chow diet was sufficient to meet the requirements of the developing brain for DHA. Alternatively, these results demonstrate that even higher dietary amounts of DHA would have had to be fed to the mothers to modify fetal brain DHA content. Feeding chow and ethanol alone decreased fetal brain PC and PE DHA content by 56.7% and 26.6% compared with controls. Feeding both ethanol and tuna with chow, however, resulted in a marked increase in the DHA concentration of both PC (66.7%) and PE (40.2%) compared both with controls and with fetuses from mothers fed ethanol and chow diet (Fig. 3). The observation that maternal supplementation with tuna oil only altered brain phospholipid DHA content in the presence of ethanol is consistent with the suggestion that ethanol may interfere with the normal mechanisms regulating PC and PE biosynthesis and turnover in the developing fetal brain. In addition to these lipid-compositional analyses, feeding tuna oil in addition to ethanol caused significant improvements to impaired motor functions of the ethanol-exposed pups. New-born guinea-pig pups from mothers fed both tuna oil and ethanol showed decreased hypotonia and a demonstrable righting reflex, although these preliminary observations were not supported by detailed physiological and neurological investigations. However, these data support tentatively the suggestion that increased availability of DHA to the fetus may ameliorate the severity of the harmful effects of ethanol on fetal neurological development and function.

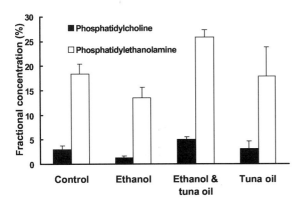

Fig. 3. Values are mean±S.D. fetal guinea-pig brain phosphatidylcholine and phosphatidylethanolamine DHA contents from pregnancies in which mothers were fed either chow, chow and ethanol, chow, ethanol and tuna oil or chow and tuna oil.

6. Summary

Although to date these observations in the fetal guinea pig have not been substantiated in humans, they provide strong preliminary evidence that impaired accumulation of DHA into brain phospholipids may be one important mechanism in the pathogenesis of FAS. Furthermore, since increasing DHA availability to the mother, and presumably the fetus, appeared to produce a reduction in the severity of ethanol-induced neurological damage it is possible that maternal DHA supplementation may provide a therapeutic strategy in humans FAS. Alternatively, the prolonged period of postnatal brain development in the human may permit appropriate dietary DHA supplementation of the affected infants after delivery, which would potentially be a more practical clinical intervention.

References

Abel, E.L. and Sokol, R.J. (1987) Incidence of fetal alcohol syndrome and economic impact of FAS-related anomalies. Drug Alcohol Depend. 19, 51–70.

Burdge, G.C. and Postle, A.D. (1994) Hepatic phospholipid molecular species in the guinea pig: adaptations to pregnancy. Lipids 29, 259–264.

Burdge, G.C. and Postle, A.D. (1995a) Phospholipid molecular species composition of developing fetal guinea pig brain. Lipids 10, 719–724.

Burdge, G.C. and Postle, A.D. (1995b) Effect of maternal ethanol consumption during pregnancy on the phospholipid molecular species composition of fetal guinea pig brain, liver and plasma. Biochim. Biophys. Acta 1256, 346–352.

Burdge, G.C., Kelly, F.J. and Postle, A.D. (1993) Mechanisms of hepatic phosphatidylcholine synthesis in the developing guinea pig: contributions of acyl remodelling and of N-methylation of phosphatidylethanolamine. Biochem. J. 290, 67–73.

Burdge, G.C., Hunt, A.N. and Postle, A.D. (1994) Mechanisms of hepatic phosphatidylcholine synthesis in the adult rat: effects of pregnancy. Biochem. J. 303, 941–947.

Burdge, G.C., Mander, A. and Postle, A.D. (1996) Hepatic and plasma phospholipid molecular species compositions in the pregnant guinea pig: effect of chronic ethanol consumption. J. Nutr. Biochem. 7, 425–430.

Burdge, G.C., Wright, S.M., Warner, J.O. and Postle, A.D. (1997) Fetal brain and liver phospholipid fatty acid composition in a guinea pig model of fetal alcohol syndrome: effect of maternal supplementation with tuna oil. J. Nutr. Biochem. 8, 438–444.

Clandinin, M.T., Chappell, J.E., Leong, S., Heim, T., Swyer, P.R. and Chance, G.W. (1980a) Interuterine fatty acid accretion rates in human brain: implications for fatty acid requirements. Early Hum. Dev. 4, 121–129.

Clandinin, M.T., Chappell, J.E., Leong, S., Heim, T., Swyer, P.R. and Chance, G.W. (1980b) Extrauterine fatty acid accretion in infant brain: implications for fatty acid requirements. Early Hum. Dev. 4, 131–138.

Corbett, R., Berthou, F., Leonard, B.E. and Menez, J.-F. (1992) The effects of chronic administration of ethanol on synaptosomal fatty acid composition: modulation by oil enriched in gamma-linolenic acid. Alcohol. 27, 11–14.

Dobbing, J. and Sands, J. (1979) Comparative aspects of the brain growth spurt. Early Hum. Dev. 3, 79–83.

Gustavsson, L. and Alling, C. (1989) Effects of chronic ethanol exposure on fatty acids of rat brain. Alcohol. 6, 139–146.

Henderson, G.I., Patwardhan, R.V., Hoyumpa, A.M. and Schenker, S. (1981) Fetal alcohol syndrome: overview of pathogenesis. Neurobehav. Toxicol. Teratol. 3, 73–80.

Hullin, F., Kim, H.-Y. and Salem, N. (1989) Analysis of aminophospholipid molecular species by high performance liquid chromatography. J. Lipid Res. 30, 1963–1975.

Jones, K.L., Smith, D.W., Ulland, C.N. and Streissguth, P. (1973) Pattern of malformation in offspring of chronic alcoholic mothers. Lancet 1, 1267–1271.

La Droitte, P., Lamboeuf, Y. and De Saint-Blanquat, G. (1984) Lipid composition of the synaptosome and erythrocyte membranes during chronic ethanol-treatment and withdrawal in the rat. Biochem. Pharmacol. 33, 614–624.

Lanting, C.I., Fidler, V., Huismaan, M., Towen, B.C.L. and Boersma, E.R. (1994) Neurological differences between 9 year old children fed breast milk or formula milk as babies. Lancet 344, 1319–1322.

Ledig, M., Megias-Megias, L. and Tholey, G. (1991) Maternal alcohol exposure before and during pregnancy: effect on development of neurons and glial cells in culture. Alcohol Alcoholism 26, 169–176.

Littleton, J.M. and John, G.R. (1977) Synaptosomal membrane lipids of mice during continual exposure to ethanol. J. Pharm. Pharmacol. 29, 579–580.

Littleton, J.M., John, G.R. and Grieves, S.J. (1979) Alterations in phospholipid composition in ethanol tolerance and dependence. Alcohol. Clin. Exp. Res. 3, 50–56.

Lucas, A., Morley, R., Cole, T.J., Lister, G. and Leeson-Payne, C. (1992) Breast milk and subsequent intelligence quotient in children born preterm. Lancet 339, 261–264.

Neuringer, M., Anderson, G.J. and Connor, W.E. (1989) The essentiality of n-3 fatty acids for the development and function of the retina and brain. Annu. Rev. Nutr. 8, 517–541.

Postle, A.D., Al, M.D., Burdge, G.C. and Hornstra, G. (1995) The composition of individual molecular species of plasma phospholipids in human pregnancy. Early Hum. Dev. 43, 47–58.

Salem Jr, N. (1989) Omega-3 fatty acids: molecular and biochemical aspects. In: G.A. Spiller and J. Scala (Eds.), New Protective Roles of Selected Nutrients, Alan R. Liss, New York, pp. 213–332.

Samborski, R.W., Ridgway, N.D. and Vance, D.E. (1990) Evidence that only newly-made phosphatidyl-ethanolamine is methylated to phosphatidylcholine and that phosphatidylethanolamine is not significantly deacylated–reacylated in rat hepatocytes. J. Biol. Chem. 265, 18322–18329.

Sinclair, A.J. and Crawford, M.A. (1972) The accumulation of arachidonate and docosahexaenoate in the developing rat brain. J. Neurochem. 19, 1753–1758.

Streissguth, A.P., Barr, H.M., Sampson, P.D., Darb, B.L. and Martin, D.C. (1989) IQ at age 4 in relation to maternal alcohol use and smoking during pregnancy. Dev. Psychobiol. 25, 3–11.

Uauy, R.D., Birch, D.G., Birch, E.E., Tyuson, J.E. and Hoffman, D.R. (1990) Effect of omega-3 fatty acids on retinal function of very low birth weight neonates. Pediatr. Res. 28, 485–492.

Weisinger, H.S., Vingrys, A.J. and Sinclair, A.J. (1996) The effect of docosahexaenoic acid on the electro-retinogram of the guinea pig. Lipids 31, 65–70.

Wilson, R. and Bell, M.V. (1993) Molecular species composition of glycerophospholipids from white matter of human brain. Lipids 28, 13–17.

Subject Index